Watson Andrews Goodyear

The Coal Mines of the Western Coast of the United States

Watson Andrews Goodyear

The Coal Mines of the Western Coast of the United States

ISBN/EAN: 9783744790697

Printed in Europe, USA, Canada, Australia, Japan

Cover: Foto ©berggeist007 / pixelio.de

More available books at **www.hansebooks.com**

THE

COAL MINES

OF THE

WESTERN COAST

OF THE

UNITED STATES.

BY

W. A. GOODYEAR,
MINING ENGINEER.

SAN FRANCISCO:
A. L. BANCROFT & COMPANY.
1877.

PREFACE.

IN writing this little book, the object which I have had in view has been not so much to discuss the geological character of the Pacific Coast coal fields as to give, what has never yet been published, a full and intelligible description of the mines themselves as they exist to-day. To what extent I have succeeded in accomplishing this, the reader must judge.

I regret that my acquaintance with the mines of British Columbia, which are for the most part confined to Vancouver's Island, is not sufficient to justify me in attempting to give any particular account of them. I have, therefore, excluded them from this work, although they are of no little importance, and are rapidly increasing their annual production.

The volume, in its present form, is mainly the result of my own work, travels and observations, extending over a period of nine or ten years, during which period it is safe to say I have done more work in, and have been personally more familiar with the actual condition and workings of, the various coal mines of the Pacific Coast than any other engineer has done.

In addition to this, however, I am also greatly indebted, as the text itself will show, to the labors of my friend and former partner, Mr. Theodore A. Blake, M.

E., who is even more intimately acquainted with the mines of Washington Territory than I have been myself, and whose early investigations of the Seattle coal field, in particular, were exceedingly thorough and valuable. I desire, furthermore, to express my obligations especially to Mr. P. B. Cornwall, President of the Black Diamond Coal Company; and, in general, to the officers and superintendents of the other coal companies throughout the country, for the liberality with which they have not only furnished all such information as I have directly asked from them, but also freely placed at my disposal every other facility for acquiring a full and thorough knowledge of their respective mines.

W. A. GOODYEAR.

SAN FRANCISCO, March, 1877.

TABLE OF CONTENTS.

	PAGE
PREFACE	3

CHAPTER I.

CALIFORNIA:
- THE MT. DIABLO COAL FIELD ... 6
  - The Clark Vein ... 12
  - The Little Vein ... 25
  - Section of Strata on Line of Clayton Tunnel ... 27
  - The Black Diamond Bed ... 30
  - Faults and Disturbances ... 36
  - Ventilation ... 44
  - Haulage, Storage and Transportation ... 48
  - Pumping and Drainage ... 52
  - Peacock and San Francisco Mines ... 55
  - Central Mine ... 56
  - Empire Mine ... 61
  - Teutonia Mine ... 64
  - Rancho de Los Meganos ... 66
- THE CORRAL HOLLOW COAL FIELD ... 69
- THE LIVERMORE MINE ... 72
- OTHER COAL LOCALITIES ... 73

CHAPTER II.

OREGON:
- THE COOS BAY MINES ... 83
  - The Eastport Mine ... 86
  - The Newport Mine ... 88
  - New Mine ... 90
- MISTAKES AND FAILURES ... 92

TABLE OF CONTENTS.

CHAPTER III.

WASHINGTON TERRITORY:

 Bellingham Bay Mine 99
 Talbot Mine 103
 Renton Mine 104
 Seattle Coal and Transportation Company's Mines 106
 Puyallup Caking Coal 128

CHAPTER IV.

MISCELLANEOUS:

 Cost of Production at Mt. Diablo Mines 130
 Statistics of Production and Trade 132
 Relative Values of Different Coals 140

CONCLUSION 151

CHAPTER I.

CALIFORNIA.

THE coal fields of the western coast of North America are limited in extent, and of comparatively recent geological origin. They are none of them of the Carboniferous Age, and, indeed, so far as yet known, none of them date back of the Cretaceous Period. They mostly furnish a non-caking, bituminous coal, which belongs to the class of lignites or brown coals. Vancouver's Island, however, produces caking coal; and some caking coal of good quality has also been found in Washington Territory. Small quantities of anthracite have been found on Queen Charlotte's Island, and probably also in Washington Territory. But no workable mine of anthracite has ever been discovered on the Coast, and the little that has been found has always proved, on investigation, to have been the result of local and special metamorphism. Of the two States and one Territory which border the Pacific Ocean between Mexico and British Columbia, Washington Territory is by far the most liberally supplied with coal. Oregon comes next, and California last. In fact, California is decidedly unfortunate in the extent and the character of her coal fields. For, although it is easy to find coal at many localities in the Coast Range from one end of the State to the other, as well as at certain points in the western foot-hills of the

Sierra Nevada; yet it generally happens either that its quality is poor, or its quantity is small, or else that it is situated in the heart of the mountains so far from market that the cost of transportation alone would far exceed the value of the coal.

I begin this treatise, however, without further prelude, by giving a description of the only field within the State where coal has hitherto been profitably mined, viz.:

The Mt. Diablo Coal Field.

The extent of the Mt. Diablo coal field may be stated in broad terms to be some ten or twelve miles along the line of outcrop of the beds running through the northern part of township 1 north, range 1 east, and the northwestern and central portions of township 1 north, range 2 east, Mt. Diablo meridian.

The details of this line of outcrop are, in many places, very irregular, and especially so in the western portion of the field, where the hills are high, and the cañons are deep and steep. But its general course may be described as follows: It is curvilinear, and convex towards the north. Beginning in the north-east quarter of section 7, township 1 north, range 1 east, it runs at first north-easterly, but curves rapidly to the east till it reaches a point in the north-west quarter of section 8, from whence it follows for almost three miles a nearly true east course across the northern portions of sections 8, 9 and 10, and close to the northern edges of these sections. But in going easterly across section 11, it bends to the south, and crossing the south half of section 12, enters the south-west quarter of section 7 in the adjoining township. From thence it follows an irregular south-easterly course across the north-western and central portions of the township as far as the "Brentwood Mines," upon the Rancho de Los Meganos, and near the line between sections 22 and 27 of township 1 north, range 2 east. Beyond this locality to the south-east, the beds have not been traced with any certainty. The dip, throughout, is in a northerly direction; but it varies in amount at different localities from 12° or 15° up to 32° or 33°, being generally highest in the western portion of the field.

A range of high hills, whose culminating points on sections 8 and 9 reach altitudes of fifteen hundred to seventeen hundred feet above the sea, runs in a general east and west direction across the northern half of township 1 north, range 1 east, and is separated by a narrow valley on the south from the still higher mountainous region which culminates in the double summit of Mt. Diablo itself, three thousand eight hundred and fifty-six feet in height.

In going east, however, from section 9, the hills diminish in height, and following the line of the coal beds south-easterly across the next township, they gradually fall lower and lower, till we reach the level of the valley at the "Brentwood Mines," which are situated in the edge of the San Joaquin plain, at an altitude of only between one hundred and two hundred feet above tide-water. In its higher portions, this range of hills is deeply scored by cañons in all directions, and it is among these cañons in the northern slopes of the range, that the hitherto paying mines are located.

The strata have been considerably disturbed at numerous localities by faults of greater or less magnitude; and the coal beds themselves are subject, within short distances, to so great variations in thickness and quality of coal, as well as in the character of the rocks which inclose them, that it is not possible with present knowledge to certainly recognize any single bed in the eastern portion of the field as being the same with any one of those which have been so extensively worked in the western portion.

By the phrase, "Mt. Diablo coal field," as here used, must be understood not merely the actually productive region, but the whole extent of the belt through which there has been found some definite evidence of probability that the beds were once continuous, or nearly so, and within which sufficient discoveries have been made to lead to the expenditure of any considerable sums of money in explorations and attempts to develop new mines. The area within which the mines have hitherto been profitably worked, however, is far more limited in extent. It lies among the higher hills in the western portion of the belt above described, and includes a distance of only about two miles and a half along the strike of the beds, from the western limits of the Black Diamond Company's workings in the north-east quarter of section 7, where the beds either split up and run out, or become too much crushed and broken to pay for working, to the most eastern limits of the Pittsburg Company's workings in the south-west quarter of section 3, and the north-west quarter of section 10, where they are stopped by the wall of a great fault which intervenes between them and Stewart's mine on the east.

The "Central (*i. e.*, Stewart's) Mine" is not here included within the profitably productive limits, for the simple reason that while it has produced considerable coal, its shipments having been sometimes as high as a thousand tons per month, it is more than probable that its production has been at a loss, instead of a profit, to its owners.

Within the productive limits above indicated, the chief openings of the mines as well as the dwellings of the miners, and other buildings, are, owing to the topography of the country, concentrated at two considerable villages, about a mile apart. The first of these villages, known as "Nortonville," is located on the south-east quarter of section 5.

The second one, known as "Somersville," is chiefly on the south-east quarter of section 4, township 1 north, range 1 east. Each village is in the bottom of a sort of amphitheatre among the hills, and at the head of a deep cañon, which runs northerly some three miles to the edge of the San Joaquin plain, from which point the distance north across the plain to the river is also in each case about three miles. Down each of these cañons there runs a railroad of the ordinary gauge (4 feet $8\frac{1}{2}$ inches) to points of shipment on the San Joaquin river, just above its junction with the Sacramento. Each railroad is therefore about six miles in length, the Black Diamond Railroad running to the Black Diamond Landing (otherwise known as New York Landing), and the railroad from Somersville (called the "Pittsburg Railroad") running to Pittsburg Landing, some two or three miles further up the river. The height of the villages themselves above tide-water ranges from seven hundred to eight hundred and fifty feet.

In the north-west corner of the south-east quarter of section 5, a round-topped hill rises to a height of one thousand three hundred and forty-eight feet above low water, and near the middle of the line between the north-west and south-west quarters of section 4, a similar hill rises to a height of about one thousand five hundred feet. Each of these two hills (between which runs the cañon of the Black Diamond Railroad) is connected with the hills to the south, in which lie the mines, by a saddle some three hundred or four hundred feet lower than its own summit—the saddle between Nortonville and Somersville being some three or four hundred feet higher than the village of Nortonville itself.

There are few points in the hills containing the mines which rise to a greater height than the higher of the two hills just described. At Nortonville, as well as at Somersville, the cañon, at the head of which stands the village, forks into numerous branches which spread upwards in all directions to the south, south-east and south-west among the hills, thus cutting up the surface of the mining ground by rough and precipitous gulches, often two hundred to three hundred feet in depth, so that the line of outcrop of the beds, as already stated, is at this locality very irregular in detail and deeply indented by the gulches. The rocks which inclose the mines consist of unaltered, grayish and reddish silicious sandstone, generally not very hard, alternating with occasional strata of rather soft clay-rock, the whole belonging to the latest formations of the cretaceous period.

The coal beds, which have been profitably worked to a greater or less extent, are three in number, and are known respectively as the "Clark Vein," the "Little Vein," and the "Black Diamond Vein." Of these, the Clark Vein is the highest in stratigraphical position; next in order below it comes the Little Vein; while the Black Diamond Vein is the lowest, and underlies both the others. The beds lie nearly parallel with each other, all dipping to the north; and at the immediate localities of the villages, both of Nortonville and Somersville, the amount of dip is from 30° to 32°.

In the Clayton tunnel, at Nortonville, the level distance from the floor of the Clark Vein south to the roof of the Black Diamond Vein is six hundred and ninety-six feet, and the dip here being about 31°, it follows that the total thickness of the strata, including the Little Vein, between the Clark and Black Diamond Veins is, at this locality, about three hundred and fifty-nine feet. At certain points the level distance between the beds is somewhat less than it is here, while in other places it is considerably greater. This is due mainly to changes in the degree of dip of the beds; though it is more than probable that the actual thickness of the strata between them also varies somewhat at different localities.

All the valuable mining ground here for a distance of nearly two miles and a half is now owned and controlled by three different companies, viz.: the Black Diamond Coal Company, the Union Coal Company, and the Pittsburg Coal Company. The Black Diamond Company owns the south half of section 5, the north half of section 8, and the north-west quarter of section 9. The Union Company owns the south-west quarter of section 4, and used to lease and mine the coal also in an adjoining strip to the east, between six hundred and seven hundred feet wide, on the south-east quarter of section 4. The Pittsburg Company owns the balance of the southeast quarter of section 4, together with an additional tract which covers portions of the north-east quarter of section 9, the north-west quarter of section 10, and the south-west quarter of section 3.

The Clark Vein.

The only bed which has been worked continuously throughout the whole distance controlled by these companies is the Clark Vein. This bed varies in thickness at different points from a minimum of eighteen or twenty inches to a maximum of four feet and a half, or a trifle over. The greatest variations in the thickness of this bed, however, do not occur within the limits of the Black Diamond Company's property, the minimum thickness in that property being twenty-eight or twenty-nine inches, and the maximum thirty-eight or thirty-nine; while the average for the whole mile of the Clark Vein controlled by this company, and as deep as the workings have yet extended, is thirty-two or thirty-three inches. On going east from the section line, however, into the south-west quarter of section 4 the bed grows rapidly thinner, and for a considerable distance in the western portion of the Union Company's ground its thickness ranges under twenty-four inches, being sometimes as low as eighteen. But in the eastern portion of the Union mine it again increases in thickness, and maintains across the south-east quarter of section 4 a thickness of from three to four feet, reaching its maximum in the Pittsburg Company's ground not far from the line between sections 3 and 4, where, as already stated, it is sometimes a little over four and a half feet thick. The Clark Vein is generally free from interstratification of slate or dirt of any kind, and with the exception of a certain portion in the western part of the Black Diamond Company's mines near the south-west corner of section 5, where, for several hundred feet it has been rather badly crushed by movements and bendings of the strata, the whole of it makes good, clean coal. Its roof and floor are also generally very good, so that it requires but little timbering. The floor is everywhere good solid sandstone, and the roof throughout the Black Diamond Company's mines, and in the western portion of the Union Company's ground, with few and small exceptions, consists also of the same material. But where the coal begins to increase in thickness in the eastern part of the Union mine, a thin stratum of rather soft clay rock makes its appearance on top of the coal and between it and the overlying sandstone. Further east, this clay stratum is nearly continuous, forming, generally, the immediate roof of the coal throughout the Pittsburg Company's ground, and seeming, as a rule, to be thickest where the coal is thickest. It reaches, in places, a maximum thickness of from two to two and a half feet, and as it separates easy from the overlying sandstone, it causes, of course, some extra trouble and expense, and occasionally more or less danger in mining the coal. But it is nothing very serious.

The chief openings to the Clark Vein are, first, the Black Diamond Company's openings, which are three. Of these the first is what is known as the "little slope," or the "hoisting slope;" the second is the "Mount Hope slope," and the third is the Black Diamond shaft. Second, the Union Company's slope. Third, the slope of the old "Eureka Company," which formerly owned a tract about eleven hundred feet wide immediately adjoining the Union Company on the east, and now belonging to the Pittsburg Company. Fourth, the Pittsburg slope. Fifth, the "Independent shaft," situated on ground formerly owned by the old "Independent Company," but now also belonging to the Pittsburg Company.

The mouth of the "hoisting slope" of the Black Diamond Company is situated in the bottom of a deep ravine which runs up southerly and south-westerly among the hills, and is eight hundred and thirty-three feet above low water in the San Joaquin river. This slope, which is ninety-eight feet long, goes down to the south with a pitch of about 35° through the strata overlying the Clark Vein. From its foot, a level gangway, known as the "Clark Vein Main Gangway," has been driven east and west on the Clark Vein throughout the company's property, and is over a mile in length across the southern part of section 5. The slope is furnished with a double track, and a steam hoisting engine whose cylinder is $14'' \times 30''$. Through this slope has been hoisted nearly all the coal which has come from the Clark Vein within the limits of the Black Diamond Company's property above the level of the Clark Vein main gangway, besides all the coal which has come from the Black Diamond Vein through the Clayton tunnel.

The mouth of the "Mount Hope slope" is situated

about four hundred and fifty feet north-easterly from that of the "hoisting slope," and is seven hundred and ninety-seven feet above low water. The slope is two hundred and ninety-three feet long to the Clark Vein, and has an inclination, or pitch, of 37° 15' to the south. From its foot, the "Mount Hope Gangway" runs east and west on the Clark Vein through the property, and is also over a mile in length. The height of the "lift" between the Mount Hope and the Clark Vein main gangways (*i. e.*, the slope distance measured on the dip of the coal bed from centre to centre of these gangways), is in the vicinity of the Mount Hope slope about three hundred and six feet. But in the western portion of the property the height of this "lift" increases considerably, owing to a decrease in the dip of the bed, the gangways being driven nearly level, or with only so much grade, about one in one hundred, as is necessary in order to drain them, and to facilitate the hauling out of the loaded cars. The Mount Hope slope is provided with a double track and a hoisting engine, with cylinder $14'' \times 30''$. From a point on the Mount Hope gangway eighty-five feet east of the foot of the Mount Hope slope, a double tracked counter-slope runs down on the coal with a pitch of about 31° to the north, a distance of three hundred and seventy-seven feet to the "Lower Mount Hope Gangway," which is the lowest gangway yet opened on the Clark Vein by the Black Diamond Company. The coal from this lower gangway, until they began to hoist through the shaft, was hoisted up the counter-slope by an underground hoisting engine $16'' \times 30''$, placed at the head of the counter-slope, and supplied with steam from the boilers at the surface by a pipe leading down the Mount Hope slope.

The Black Diamond shaft is distant about six hundred and twenty feet in a direction a little north of west from the mouth of the Mount Hope slope. It is a vertical shaft, heavily and well timbered, and measures twenty-two feet four inches by eleven feet ten inches from out to out, being divided into three compartments, viz.: one pumping compartment five by nine feet, and two hoisting compartments, each six by nine feet in the clear, inside of timbers. The mouth of the shaft is eight hundred and thirty-nine feet above low water, and its present depth to the level of the lower Mount Hope gangway is four hundred and fifteen feet, and the foot of the shaft is about fifty feet south of this gangway. The shaft is furnished with two iron, double-decked safety cages, each cage raising two loaded mine cars at a time, and each car containing about a ton of coal. The hoisting power for this shaft consists of a pair of large steam engines working directly on the winding shaft, each engine having a twenty-four inch cylinder, and five foot stroke. The cables used here are flat wire ropes winding on spools.

The mouth of the Union Company's slope is situated very close to the line between the south-east and south-west quarters of section 4, and is eight hundred and sixty-six feet above low water. The slope itself is four hundred and seventeen feet long to the Clark Vein, with a pitch of 37° 45′ to the south. From its foot a gangway runs east and west on the Clark Vein through this company's property. From a point on this gangway two hundred and forty-four feet west of the foot of this slope, a counter-slope runs down on the bed with a pitch of 28° 23′ to the north, three hundred and four feet to a second gangway, and then about three hundred feet further to the third or lowest gangway in this mine. Each of these slopes was worked by a steam hoisting engine placed at its head, the underground engine at the head of the counter-slope being supplied with steam from boilers at the surface as in the case of the Mt. Hope and its counter-slope.

The old Eureka slope was about two hundred and ninety feet long, with an average pitch of about 43° 15' to the south. It was furnished with a three-rail track and a hoisting engine, 10"×24". Its mouth is seven hundred and eighty-six feet above low water. There was also a counter-slope fifty-five feet west of the foot of the surface-slope which went down some six hundred feet or more, and was furnished with a double track and a hoisting engine, 12"×24". The whole of that portion of the Clark Vein originally owned by the Eureka Company has already, however, been worked out and exhausted, and these openings are therefore now abandoned.

The Pittsburg slope is in the southeast corner of section 4. Its mouth is eight hundred and thirty-eight feet above low water. It goes down in a direction somewhat to the west of south, with a pitch of 25° 50', and is about two hundred and forty feet long to the Clark Vein. It is furnished with double track and steam hoisting engine, 12"×24". From its foot a gangway runs both ways on the bed through the company's property. From a point on this gangway twenty-five feet west of the foot of the surface-slope, a counter-slope runs down on the dip about eight hundred feet with a pitch of about 31° 30' to the lowest gangway in the mine. There are, however, two intermediate gangways, one at a point three hundred feet and the other at a point five hundred and seventy-nine feet down from the head of the counter-slope. The slope is double-tracked and is worked by a steam engine, 14"×30", at its head.

In the eastern part of this mine and distant nearly a a quarter of a mile from the foot of the surface-slope, there is another counter-slope running down from the upper gangway to the second one. This slope is also double-tracked and furnished with a hoisting engine. Both the underground engines are furnished with steam through pipes conducting it from the boilers at the mouth of the surface-slope.

The Independent shaft is a vertical shaft sunk by the now defunct Independent Company about the year 1865, at a point a little south-west of the centre of the north-east quarter of the south-east quarter of section 4. The size of this shaft from out to out is nearly sixteen feet by ten, and it is divided inside the timbers into three compartments, viz.: one pumping compartment, three feet by seven feet eight inches, and two hoisting compartments, each four feet by seven feet eight inches in the clear. Its mouth is seven hundred and nineteen feet above low water, and it is seven hundred and ten feet deep.

Some curious engineering was displayed in connection with the sinking of this shaft. It was represented, for example, at the time of its commencement, that it would strike the Clark Vein at a depth of from four hundred to four hundred and fifty feet; yet the position of the Clark Vein, and the amount of its dip in this vicinity, were at that time already well known, and a little simple measurement and computation would have demonstrated then as well as now the fact that a vertical shaft to the Clark Vein at that locality, must be in the neighborhood of nine hundred and fifty feet in depth instead of four hundred or four hundred and fifty. However, they went down seven hundred and ten feet, and then getting tired of sinking, left about twenty-four feet in the bottom of the shaft for a sump, and started a level tunnel to the south, at the depth of six hundred and eighty-six feet, for the coal. They drove this tunnel to the Clark Vein, its length proving to be (according to the best accounts afterwards obtainable) about four hundred and twenty feet. When within about one hundred and fifty feet of the coal, they struck a large stream of water, which necessitated heavy pumping machinery, and a steam engine was erected and a Cornish pump put in. But the foundations for the engine were bad, and it was never firm upon its bed. Moreover, the only available water was that from the mine, which was so heavily charged with a mixture of various sulphates, with probably some free sulphuric acid, as to be exceedingly destructive to the boilers. Nevertheless, the Clark Vein was at last reached, and a gangway driven, and a considerable area of coal worked out above its level, stretching upward towards the lower workings of the Eureka Company. But the work was done at a heavy loss, and it was finally abandoned, the parties who inaugurated it becoming bankrupt. It is said that the amount of money expended here before the coal was reached was over one hundred and fifty thousand dollars. While the work was in progress a connection was made with the lower

Eureka workings by a shoot driven up along the coal for purposes of ventilation. This shoot afterwards answered the purpose of a water-drain for the Eureka mine; and the Eureka Company, having in the course of time purchased the Independent property, employed the Independent shaft, at considerable expense, for a year or two, merely as a pumping shaft, in which the water was held at a certain level, which just sufficed to drain the lower workings of their own mine. But after having exhausted the Clark Vein in their own property down to the top of the old Independent workings, the Eureka Company, in its turn, stopped work, and subsequently sold both the Eureka and Independent properties to the Pittsburg Company, which now owns them. The Independent shaft is now abandoned and idle. It will be noticed, however, that the level at which the tunnel from the foot of this shaft struck the Clark Vein is at a considerably greater depth than any other point which has yet been reached in the Mt. Diablo mines.

As a general rule, the coal close to the surface of the ground is not of good quality in any of the beds. The outcrop of the Clark Vein is in many places nothing more than a soft clay shale, light brown in color, and very slightly carbonaceous. At a few points there was good coal in the Clark vein very close to the surface; but the depth at which the coal in this bed first becomes marketable generally ranges from one hundred to two or three hundred feet, measured on the dip of the bed.

The height to which the coal has been worked in the hills up towards the outcrop, varies greatly, of course, at different localities, depending on the configuration of the surface, which the ravines have scored so deeply;

and at one point depending on another circumstance, viz., the fact that immediately to the south of the "Little Slope" of the Black Diamond Company, and extending for a considerable distance, both east and west of it, there was an area of several acres in which the coal was entirely wanting, and which presented every appearance of having been at some time in the past on fire and burned out. The maximum height, however, to which the coal has been mined on the Clark bed, within the limits of the Black Diamond Company's property above the Clark Vein main gangway, is about six hundred and seventy-five feet; and a fair statement for the *average* height of the workings above this gangway for the whole distance of a mile across the southern part of section 5, would be about five hundred feet.

The height of the "lift," also about one mile in length, between the Mt. Hope, and the Clark Vein main gangways varies from a minimum of three hundred and six feet in the eastern portion of the mine to a maximum of about five hundred in the western portion, owing to a decrease in the dip of the bed going west. Probably a fair average for the height of this lift for the whole mile would be about three hundred and fifty feet. The height of the next lower "lift," and the present lowest one on the Clark Vein, in the Black Diamond Company's mines, *i. e.*, the lift between the Mt. Hope and the Lower Mt. Hope gangways, increases going west from three hundred and seventy-seven feet in the eastern part of the mine to about four hundred and fifty in the western part. This "lift" is now only about four thousand three hundred feet long, the Lower Mt. Hope gangway not having been driven so far to the west as the upper gangways have been.

The Clark Vein at the present time (1877), may be said to be practically exhausted throughout the Black Diamond company's property down to the level of the Lower Mt. Hope gangway, the quantity of coal which yet remains to be extracted from above that level being very small; but below this gangway it is all untouched and solid.

In the Union mine, they have worked above their Clark Vein gangway No. 1 (*i. e.*), the gangway at the foot of their surface-slope, for a height of about 600 feet up to the old Manhattan gangway. The latter gangway (driven by the old Manhattan Company, which formerly owned the north-west quarter of section 9, and afterwards sold it to the Black Diamond Company), started on the outcrop of the Clark Vein in the bed of a gulch in the south-west part of the south-east quarter of section 4, and was driven westerly on the Clark Vein some twenty-four hundred or twenty-five hundred feet, extending for most of this distance into and along the southern edge of the south-west quarter of section 4. Some coal was extracted here by the Manhattan Company, and their workings are said to have extended in places as high as about three hundred feet above this gangway. But these old workings have long been caved and closed, and there is no very reliable information obtainable now as to their exact extent. The Manhattan Company also drove a tunnel from this gangway south to the Black Diamond Vein, and opened several hundred feet of a gangway on that bed, but never took out much coal from there.

The height of the lift in the Union Mine between the Clark Vein gangway No 1 and the Clark Vein gangway No. 2 is about three hundred and four feet; and throughout the Union Mine the Clark Vein is now exhausted down to the level of this latter gangway, which is only a few feet higher than the lower Mt. Hope gangway of the Black Diamond Company. Furthermore, the Union Company have continued their counter-slope some three hundred feet further down the dip, and driven from its foot another gangway (their Clark Vein gangway No 3), from which, in 1876, they were working an additional lift of about three hundred feet of coal. This gangway was driven west only about eighteen hundred feet from the foot of the counter-slope when the mine was abandoned and closed in December, 1876.

There are no accurate records now obtainable of the extent of the workings in the upper part of the old Eureka Mine. But the Clark Vein was entirely cleaned out here as far up as it would pay to work towards the outcrop, and it is evident, from the situation of the slope and the shape of the hills at that locality, that these workings must have extended in places to a height of some five or six hundred feet above the gangway which runs at the foot of their surface-slope. The Eureka Company also worked out through their counter-slope two additional lifts, aggregating some six hundred feet or more below this gangway, and cleaning out everything down to the top of the old Independent workings, from which point the distance down through the latter to the gangway at the level of the tunnel from the foot of the Independent shaft, must have been in the neighborhood of four hundred feet, since the total distance measured on the dip of the bed from this lowest gangway up to the gangway at the foot of the Eureka surface-slope was between one thousand and eleven hundred feet. Thus, for a certain distance, here in the ground formerly owned by the Eureka and Independent companies, the Clark Vein is practically exhausted to the depth of fifteen hundred or sixteen hundred feet measured on the dip of the bed from the outcrop down to the tunnel from the foot of the Independent shaft. But a direct comparison of this distance with the distances down from the outcrop at other localities, will not give the correct differences of absolute height between the bottoms of the mines, since the hills along the line of outcrop vary a good deal in height, and rise considerably higher in the Union and Black Diamond properties than they do here in the old Eureka. In fact the distance on the dip of the bed from this lowest gangway of the old Independent Company up to the level of lower Mt. Hope gangway of the Black Diamond Company, would be about seven hundred and forty feet.

In the Pittsburg mine, the coal on the Clark Vein has been worked to heights ranging from six hundred feet, or less, to a maximum of between nine hundred and a thousand feet on the dip above the gangway which runs at the foot of their surface-slope, and above the highest of these workings the additional distance up, through soft and worthless coal and shale to the outcrop itself, is from two hundred to three hundred feet. Below the foot of the surface-slope, there have been worked through the counter-slope in the Pittsburg mine three additional lifts of three hundred, two hundred and seventy-nine, and two hundred and twenty-one feet respectively, making the total maximum depth worked on the Clark Vein in this mine, measured on the dip, between seventeen hundred and eighteen hundred feet; and below the lowest Pittsburg gangway it will still be about five hundred and thirty feet farther down the dip to the level of the tunnel from the foot of the Independent shaft.

The Little Vein.

Before speaking of the workings on the "Little Vein," it will be well for the sake of clearness to describe two tunnels in the Black Diamond Company's property which are driven from daylight south into the hills, and run nearly level through the superincumbent strata to the Black Diamond Vein. The first of these is known as the Upper Black Diamond Tunnel. This is the highest and the oldest of all existing openings to the Black Diamond Vein. Its mouth is in the right hand or south-east side, and near the bottom of the same ravine in which is situated lower down the mouth of the Clark Vein hoisting slope. It is distant from the latter a little over eight hundred feet, in a direction somewhat to the west of south, and is one thousand and thirty-four feet above low water level. It is also some little distance to the south of the line of outcrop of the Clark Vein. The tunnel is straight, and runs in a direction about 2° east of true south for a distance of four hundred and twenty-two feet to the Black Diamond Vein. At a point one hundred and twenty feet from its mouth, it cuts a seam of coal, which is at this locality fourteen inches thick. At a point one hundred and ninety-six feet from its mouth it cuts another and smaller seam. These two are the only seams of coal exposed in the Upper Black Diamond Tunnel above the Black Diamond Vein; the rest of the strata consisting of sandstones and shales.

The second of the two tunnels now in question has been already mentioned, and is known as the Clayton Tunnel. Its mouth is in the bottom of the ravine, about one hundred and sixty feet south-westerly from the mouth of the Mt. Hope slope, and is seven hundred and ninety feet above low water. This tunnel is also straight, and runs about $4\frac{1}{2}°$ to the west of south, a distance of nearly eleven hundred feet to the Black Diamond Vein. It has an ascending grade of about one foot in a hundred going in. The distance in from its mouth to the point where it cuts the centre of the Clark Vein is three hundred and sixty feet.

The following sketch shows a section of the strata as exhibited along the line of this tunnel. The various strata are numbered in the sketch to correspond with the numbers in the first column of the tabular description which immediately follows. The dip here being about $31°$, the actual thickness of the strata is a trifle over one half the distances which they occupy respectively along the level floor of the tunnel. In the last two columns of the tabular description, both these figures are given; the first one containing the respective level distances along the floor of the tunnel, and the last one the actual thicknesses of the various strata in feet and decimals of a foot.

CALIFORNIA.

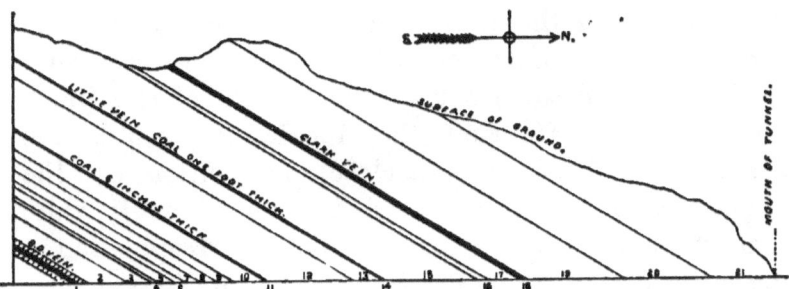

SECTION OF STRATA ON LINE OF CLAYTON TUNNEL.

DESCRIPTION.	Distance on floor of tunnel, in feet.	Actual thickness of strata, in feet.
1. BLACK DIAMOND VEIN, forty inches coal, with seven or eight feet of bone above and below..............................	35.0	18.0
2. Heavy-bedded sandstone—Dry..	66.5	34.2
3. Heavy-bedded sandstone—Wet......	40.0	20.6
4. Thin-bedded and ferruginous sandstone.....................	4.0	2.0
5. Heavy-bedded sandstone—Wet............................,......	31.0	16.0
6. Thin-bedded carbonaceous shales—Dry.....................	10.0	5.2
7. Numerous alternations of sandstones and thin shales; some of the latter carbonaceous—Dry............................	35.0	18.1
8. Heavy-bedded sandstone—Dry..	20.0	10.3
9. Alternations of sandstone and thin shales, the latter sometimes carbonaceous—Wet......	25.0	12.9
10. Same alternations as in 9, but dry............................	45.0	23.2
11. COAL, five or six inches thick................................	1.0	0.5
12. Alternations of sandstones and thin shales—Dry............	139.0	71.6
13. Heavy-bedded sand-rock—Dry....................................	48.0	24.7
14. COAL about one foot thick, with six inches of bone on each side of it......................................	4.0	2.0
15. Heavy-bedded sand-rock—Dry....................................	146.0	75.2
16. Heavy-bedded sand-rock—Dry....................................	18.0	9.3
17. Coarse-grained sand-rock, generally heavy-bedded............	64.0	33.0
18. CLARK VEIN, thirty-four inches of coal................	5.5	2.8
19. Same sandstone as in 17..	137.5	70.8
20. Fine-grained, bluish and clayey rock, moderately heavy-bedded, with occasional bands of coarser sand-rock a few inches thick........	127.0	65.4
21. Thin-bedded clay shales, to mouth of tunnel................	93.0	47.9

Between 15 and 16 in the above section there is a streak of carbonaceous shale one foot thick, and also a second one of about the same thickness between 16 and 17.

As there are two little seams of coal above the Black Diamond Vein in the Upper Black Diamond tunnel, so also there are two little seams in the Clayton tunnel between the Black Diamond Vein and the Clark Vein. But it will be seen, on closer inspection, that the positions of the upper and lower little seams in the two cases do not correspond to each other respectively. In fact, their distances from the middle of the Black Diamond Vien are such that the lower seam in the Clayton tunnel corresponds closely in position with the upper seam in the upper Black Diamond tunnel; while there is no seam in the Clayton tunnel corresponding to the lower one in the upper Black Diamond tunnel; and furthermore, the mouth itself of the latter tunnel appears to be further south by some sixty feet than the proper position at this level for the upper seam of the Clayton tunnel. Therefore, as the distance between these two tunnels in the direction of strike of the beds is only some three or four hundred feet, and as no faults of any magnitude have been discovered within this distance in any of the workings on either the Clark or the Black Diamond Vein, it is extremely probable that the lower seam in the Clayton tunnel is really identical with the upper one in the upper Black Diamond tunnel, in spite of the fact that if we considered the respective thicknesses of the seams alone we might be induced to draw an opposite inference. Moreover, the entire disappearance of the lower little seam of the upper Black Diamond tunnel within the short distance of less than four hundred feet between it and the Clayton tunnel, and the decrease in thickness of the upper little seam of the upper tunnel within the same distance, from fourteen inches to five or six inches only, are in perfect keeping with numerous other cases throughout the Mt. Diablo mines, in which considerable variations in thicknesses of strata within short distances are matters not of doubt but of certainty, having been fully exposed by the underground workings.

It has been previously stated (page 10), that the number of beds which have been profitably worked at the Mount Diablo mines is three; but it will be seen presently that there is a strong probability that at different localities two separate and distinct beds have been confounded under the general name of "Little Vein," and that the whole number of beds which have been more or less worked with profit, should, therefore, be stated at four.

There have been no "Little Vein" workings in the property of the Black Diamond Company, for the reason that in this portion of the field none of the little seams where cut by tunnels between the Clark and Black Diamond veins have been of sufficient thickness to pay for working. But in the eastern part of the Union Mine, and also in the western part of the old Eureka Company's ground, one of these little seams reached a thickness ranging from sixteen to twenty-four inches of good coal, and has therefore been quite extensively worked, yielding an aggregate of perhaps from forty to fifty thousand tons of coal. This vein is in all probability identical with the upper little seam in the Clayton tunnel; for the total thickness of the strata between it and the Clark Vein in the Clayton tunnel is about one hundred and twenty feet, while the thickness between the Clark Vein and the "Little Vein," which has been worked in the Union and Eureka, as shown by the workings at two different localities, is in the Union ground about one hundred and seventeen feet, and in the Eureka ground about one hundred and nineteen feet. There can be no question about the Little Vein in the Eureka ground being identical with that in the Union; for the workings have actually connected with each other under-ground. But at a point considerably farther east, in the Pittsburg Mine, and in the north-east corner of the north-east quarter of section 9, a "Little Vein" has been worked to a considerable extent by means of a little slope driven down to it through the Clark Vein, at a point a few hundred feet south-west of the mouth of the main Pittsburg slope already described. And here the thickness of strata as exposed in the little slope below the Clark Vein, and above the Little Vein, is about two hundred and fifteen feet. It is probable, therefore, that this "Little Vein" is a different seam from the one in the Union and Eureka grounds, and that it underlies the latter.

THE BLACK DIAMOND BED.

The chief openings to the Black Diamond Bed, with the exception of the two tunnels already described, do not run directly out to day-light, but are in the shape of tunnels entirely underground, driven south from various gangways on the Clark Vein.

The Black Diamond Company has four such tunnels. Two of these run south directly from the Black Diamond shaft. The first or upper one, known as the "Black Diamond Tunnel No. 2," is at the level of the Mount Hope gangway, which it intersects at a point two hundred and sixty feet south of the centre of the shaft. The second, and lower one, known as the "Black Diamond Tunnel No. 3," starts from the foot of the shaft at a point about six feet higher than the level of the lower Mount Hope gangway, so that the cars from this tunnel will run on to the upper platform of the double-decked cage which receives at the same time on its lower platform the cars from the lower Mount Hope gangway.

From the south ends of these tunnels, gangways are driven both east and west on the Black Diamond Vein, and are known respectively as the "Black Diamond Gangway No. 2," and the "Black Diamond Gangway No. 3." The gangway described hereafter at the south end of the Clayton tunnel, and formerly known as the "Lower Black Diamond Gangway," being now called the "Black Diamond Gangway No. 1."

About twenty-one hundred to twenty-two hundred feet west from the shaft, two other tunnels are driven south to the Black Diamond Vein, one from the Mount Hope, and the other from the lower Mount Hope gangway. These tunnels are known respectively as the "West Black Diamond Tunnel No. 2," and the "West Black Diamond Tunnel No. 3."

In the Union Mine, a tunnel runs south from the Clark Vein gangway at the foot of the surface slope, and in line with that slope to the Black Diamond Vein; and a lower tunnel to the same vein runs south from the gangway on the Clark Vein next below the one at the foot of the surface slope, and is about seventy-five feet east of the counter-slope already described in this mine. From the south end of the upper one of these tunnels a gangway was driven a few hundred feet to the west; but the driving of this gangway was stopped when it was found that it was south of the section line, and therefore in ground belonging to the Black Diamond Company. From the south end of the lower tunnel, a gangway has been driven a long distance west, and a considerable lift of coal worked out above it and between it and the section line.

From the south end of the upper Black Diamond tunnel already described, in the ground of the Black Diamond Company, a gangway was driven a few hundred feet to the east, and over three-fourths of a mile to the west.

From the south end of the Clayton tunnel the Black Diamond Gangway No. 1. (formerly known as the "Lower Black Diamond Gangway") was driven west considerably over three-fourths of a mile, and east a little less than three-fourths, making the total length of this gangway something over a mile and a half. This is the longest single continuous gangway which exists in the Mt. Diablo mines. For a short distance at its eastern end, it runs along but a few feet below the level of the old Manhattan Gangway on this bed, already mentioned, and driven many years ago; and this part of the Black Diamond Gangway, No. 1. was only driven for the sake of making a connection with the old Manhattan tunnel for purposes of ventilation.

The Black Diamond Vein is everywhere a much more expensive bed to work than the Clark Vein. This is owing to the bad character of the immediate roof and floor of the coal. The whole thickness of the Black Diamond bed varies in different localities from six or eight to eighteen or twenty feet. But the greater portion of this thickness consists of interstratified clay-slate, and "bone,"—the last word being a miners' term to designate a very impure, slaty and worthless coal, which forms a weak roof and a bad floor, requiring much timbering and gradually swelling so badly on exposure to the air as to crush the timbers, and necessitate frequent cutting down of the bottoms of the shoots and the gangway floors. The workable coal, wherever it extends in the Black Diamond Vein, lies nearly in the middle of the mass forming the thick bed just described, and has bone and shale both above and below it. It generally attains its maximum thickness at those localities where the whole bed reaches its maximum development, or in other words, where the workable coal is thickest; there, also, the "bone" and slate are thickest, both above and beneath it, and *vice versa*, where the total thickness of the bed is least, there the workable coal thins out or even disappears entirely and the whole bed becomes worthless. The coal itself, however, in this bed, wherever thick enough to be worked with profit, is generally clean and free from interstratified slate or "bone", and there have been considerable areas in the Black Diamond Vein which have yielded rather better, because harder, coal than most of that produced by the Clark Vein.

Throughout the whole length of the upper Black Diamond Gangway, except for a little distance in the extreme western portion of the mine, the coal was good, and its thickness averaged about forty-four inches though varying at different points from thirty-six to fifty-four.

The maximum height of the lift worked out above this gangway was about five hundred and seventy-five feet. But its average height was much less than this, amounting to not far from three hundred and fifty feet.

The height of the lift between the upper Black Diamond Gangway and the Black Diamond Gangway No. 1., varied from a minimum of about four hundred and twenty feet at a point some four or five hundred feet west of the upper Black Diamond tunnel, to a maximum of a little over five hundred and fifty feet in the western part of the mine. This increase in the height of the lift towards the west is due here, as well as on the Clark bed, to a decrease in the amount of dip as we go west.

There was a small patch of good coal left unworked above the top of the highest workings above the upper Black Diamond Gangway, which is now inaccessible from below, but which may possibly be worked out hereafter by means of a new and shallow opening from the surface of the hills: and there are also a few acres of good coal not yet worked out in the northwest quarter of section 9, above the level of the Black Diamond Gangway No. 1. But with these two exceptions, the Black Diamond Vein is already exhausted down to the level of the latter gangway.

The height of the lift between the Black Diamond Gangway No. 1 and the Black Diamond Gangway No. 2, at a point about five hundred and ninety feet east of the Black Diamond Tunnel No. 2, is three hundred and fifteen feet. At the south end of the Black Diamond Tunnel No. 3, the height of the lift from there up to the Black Diamond Gangway No. 2, is three hundred and eighty-two feet.

From the south end of the Clayton tunnel, westerly, along the Black Diamond Gangway No. 1, the coal was worked continuously all the way, and the whole lift exhausted as far west as the gangway was driven. For a few hundred feet, however, to the west of the tunnel, the coal was not quite so thick as it was in the upper gangway, the average for the first eight hundred feet here being only about thirty-four inches; and at a point one thousand and sixty feet west of the tunnel there was a small fault, beyond which, for a distance of some three hundred feet, the coal was rather soft. But west of this the coal again was good and hard, and with an average thickness of probably forty inches, reaching in places a maximum of four feet and a half.

To the east of the Clayton tunnel, the coal on this gangway was thinner, but maintained an average (though gradually diminishing), thickness of twenty-nine inches for a distance of some three hundred and thirty feet, becoming gradually, however, more and more streaked with "bone." Within the next fifty feet, it dwindled rapidly to only eight or ten inches in thickness, becoming entirely worthless, and at the same time so dirty as to be called no longer "coal," but "bone." From thence, eastward, for a distance of between eight hundred and nine hundred feet along this gangway, there was no coal of any value whatever, and the total thickness of the bed for a portion of the way was only some five or six feet, consisting entirely of slate and bone. But at a point about twelve hundred and fifty feet east of the tunnel, the coal again comes in, and then continues of fair quality and with a thickness ranging from two and a half to three and a half feet, for a distance of something like two thousand feet in the north-west quarter of section 9. This quarter-section covers a massive hill, which rises to a height of a little over fifteen hundred feet above tide-water; and it is probable that the good coal in this hill, at about the central part of the quarter-section, extends to a height of nine hundred or one thousand feet on the dip of the bed above the present gangway. But it has yet been worked to a height of only about three hundred feet above the gangway; though at one point a shoot was driven up to test its quality some seven hundred feet above the gangway, and it was found to be good as far up as this shoot extended.

At a considerable distance to the east of the old Manhattan tunnel, two or three other openings have been made to the Black Diamond Vein by means of tunnels and slopes driven by the old Eureka and the Pittsburg companies in ground now belonging to the latter company, but the coal was not found thick enough and good enough to pay for working, and no mining of any account has been done there. It is only within the mile and a half already described, which is traversed from end to end by the Black Diamond Gangway No. 1, that the Black Diamond Vein has ever been worked with any profit.

FAULTS AND DISTURBANCES.

Throughout the Mt. Diablo coal mines the beds are frequently more or less disturbed by faults and dislocations. Within the two and a half miles of profitable working, some seven or eight of these faults are of considerable magnitude, involving throws of from ten or fifteen feet to one hundred and fifty feet or more, and immediately outside of this two miles and a half, both on the east and on the west, there are disturbances of still greater magnitude. But besides these larger faults, the smaller disturbances scattered throughout the mines and involving well-marked dislocations, or throws, of from five or six feet down to as many inches or less, are extremely numerous. These disturbances are generally most sharply defined and may be most easily studied in the Clark Vein. Many of the smaller ones are entirely local in character, and extend but very short distances; and it is only a very few of the largest ones which appear to extend through the whole mass of strata between the Clark and Black Diamond Veins with sufficient uniformity in character and direction to render it possible to recognize with certainty the same fault in both the veins.

The longest distance which occurs anywhere in the mines without any fault or disturbance of noticeable magnitude, is a distance of about two thousand feet on the Clark Vein, stretching east from the Black Diamond shaft into the western portion of the Union mine. The foot of the Black Diamond shaft itself, however, is immediately opposite the point where a fault, involving a throw of from fifteen to twenty feet down to the west, crosses the Lower Mt. Hope gangway. This fault, like most of the larger ones in the Mt. Diablo mines, has a north-easterly and south-westerly course, and its plane dips at a steep angle towards the north-west. It shows through all the upper works on the Clark Vein, and its position may be traced upon the mine-map by the sudden changes in the directions of the three successive gangways on that vein, where it crosses them. It is also probable, through not certain, that this fault extends through the intervening strata to the Black Diamond Vein, as there is a fault of several feet down to west in the latter vein, which crosses the Upper Black Diamond gangway eight hundred feet west of the upper tunnel, and the Black Diamond gangway No. 1, at a point one thousand and sixty feet west of the Clayton tunnel, running north-easterly and south-westerly, and in such a position that although not exactly in line with this fault upon the Clark Vein, yet by curving slightly, as it is very likely to do, through the intermediate strata, it may very easily connect with it.

To the west of this, for a distance of over three thousand five hundred feet, there is no single fault upon the Clark Vein which equals it in magnitude, though there are many smaller ones.

In the Union mine, there are five large faults. One of these is to the east of the Union slope; one is just west of it, and is a considerable downthrow to the west; the three others are still further west, two of them being upthrows to the west, and one, the largest of all of them, a downthrow of sixty or seventy feet to the west. Some of these faults, doubtless, also run through to the Black Diamond Vein; but they show themselves there in such modified forms that among the multiplicity of minor disturbances it is not easy to recognize them.

In the eastern part of the old Eureka Company's ground there is a large fault which, so long as this company continued to work, formed practically the dividing line between its underground workings and those of the Pittsburg Company.

At a point in the eastern part of the Pittsburg Mine, about nine hundred feet east of the foot of the Pittsburg slope, a fault crosses the gangway larger than any of the preceding, and consists of an upthrow of something over one hundred and fifty feet to the east; and at the extreme eastern limit of the Pittsburg Company's workings, the gangway terminates at the wall of another fault, which has never yet been thoroughly explored, and perhaps never will be, but which, judging from the position of the outcrop of the Clark Vein, at points further east toward Stewart's mine, must consist of an upthrow to the east of not less than three hundred to four hundred feet, and possibly more.

It would be both impracticable and useless to describe all the smaller disturbances scattered through these mines. But, as an illustration of the frequency with which they sometimes occur, the following description is given from actual measurement of a somewhat remarkable "trouble" which extends for some distance along the Clark Vein main gangway, in the western part of the Black Diamond company's mines: Beginning at a point about twenty-one hundred and ten feet west of the foot of the Little Hoisting Slope, we have first a jump of seven inches down to west; then

20	feet further west, a jump of			17	inches up to west.			
7½	"	"	"	"	6	"	"	"	"
12½	"	"	"	"	12	" down	"	"
2	"	"	"	"	12	" up	"	"
6½	"	"	"	"	22	" down	"	"
34½	"	"	"	"	7	" up	"	"
2	"	"	"	"	16	"	"	"	"
10	"	"	"	"	12	"	"	"	"

Then for a short distance the ground is *very* irregular and the coal entirely disappears, with the exception of a thin and irregular seam, which bends first up and then down to the west, to where the coal comes in on the gangway again, which it does with a jump down to west at a point about seventeen feet west of the twelve-inch jump last noted. We then have

5 ft.	farther west,	a jump of		10	inches	down	to west.
4 "	"	"	"	29	"	up	"
12 "	"	"	"	4	"	down	"
2 "	"	"	"	4	"	up	"
3 "	"	"	"	2	"	"	"
2 "	"	"	"	5	"	down	"
3 "	"	"	"	31	"	"	"
14 "	"	" { an irregular roll and jump resulting in a change of }	30	"	up	"
8 "	"	"	a jump of	31	"	"	"
22 "	"	"	"	42	"	down	"

This was the last jump measured; but there were more of them, and the "trouble" extended some little distance further west. Throughout its whole extent, both coal and sandstone were of course very badly crushed, some of the latter being just ready to fall to powder and run like loose sand, and very little coal was obtained from here till, on going farther west, the faults became less frequent.

It sometimes happens that the same fault in different portions of its course varies considerably in the amount of its throw, showing that the displacement in such cases has involved a twisting of the strata. There is one notable instance of this kind in the eastern part of the Union mine. In the upper workings here a fault of considerable magnitude runs about north-east and south-west, and at the highest point of the workings exhibits itself as an upthrow of sixteen feet to the east. But it curves gradually to the north in going down, thus being convex to the east, while at the same time the amount of its throw gradually diminishes, until, at the point where it crosses the present lowest gangway in the mine, its course is about north and south, and its upthrow is only eighteen inches to the east.

With reference to the direction of throw in the faults, the general law holds pretty well throughout these mines, that, where the plane of a fault is inclined from the vertical, it is the hanging wall of the fault which has gone down. But this law, though general, is not universal, and cases are occasionally found here in which the throw is in the opposite direction.

The general line of strike of the beds, in spite of all faults and disturbances, is very straight for a distance of nearly a mile and a half in a direction about N. 86° W (true course), from the Pittsburg slope to a point about as far west as the middle of section 5; and within this distance the dip does not vary greatly from 30°, ranging in general from 28° to 32°.

But, going west from the middle line of sections 5 and 8, the beds and the strata *curve* far around in a gradual sweep to the south, while at the same time their dip gradully diminishes until it does not exceed 20°; and in the western part of the Clark Vein main gangway there were places where it was only 15°. The general form and shape of the beds as they lie in this part of the mines, therefore, is that of *warped surfaces*. And this state of things produces, of course, a gradual divergence of the gangways of each bed from each other, and a gradual increase in the height of all the "lifts" in going west.

This great curve of the beds to the south, in the western part of the mines, is evidently preliminary to a great and sudden disturbance of the strata, which is proven by other evidence to exist in the eastern halves of sections 6 and 7, and within a short distance to the west of where the mines have stopped. But none of the gangways, on either the Clark or Black Diamond Vein, were driven far enough west to actually encounter this great disturbance. They were stopped because, for various reasons, it was not profitable to drive them further. In the upper Black Diamond gangway, the coal still kept its place and thickness, but had grown rather soft and friable, and in this expensive bed, with the disadvantages of a dip of only 18° to 20°, and nearly a mile of underground haulage, it would not pay to go further for coal of so poor a quality. The face of the Black Diamond Gangway No. 1, struck a fault, the magnitude of which was not known; and though the coal here was of good quality, and over four feet thick up to the fault, yet the dip here, as well as in the upper gangway, was low, and the underground haulage was over a mile. So it was not considered advisable to drive further, upon the chances, in the face of the additional fact that the near proximity, though not the exact locality, of great disturbance to the west, was certain. In the Clark Vein main gangway, the work was finally stopped because of a somewhat interesting fact of altogether a different kind. Here the Clark Vein gradually split into two portions which grew thinner and thinner until they almost disappeared. At a point, probably a thousand feet back from the final face of the gangway, an almost imperceptible seam or parting first made its appearance in the middle of the bed. This parting at first was not thicker than a knife-blade, and it ran a considerable distance before it presented any further special change. But then it begun slowly to increase in thickness, and gradually developed itself into a little layer of clay-slate. This change went on slowly for some distance, the coal above and below the slate being still good, but decreasing slightly in thickness, till at a point four hundred and eighty-five feet from the final end of the gangway, there were two streaks of coal, each about one foot thick, with a few inches in thickness of slate between them.

Here the gangway struck the first jump of a series of small faults and rolls which continued for some distance, and there was a sudden increase in the thickness of slate between the two bands of coal. The gangway was still driven on, however, in the hope that the coal might come in again; — but with prospects which only grew worse and worse; the coal growing thinner and the slate growing thicker, till at last the upper seam of coal was only about three inches thick, and the lower one six inches, while the clay-rock between them had increased to a thickness of about five feet. The work was then abandoned.

The exact condition of the face of the Mt. Hope gangway when abandoned is not known to the writer. But it is evident in any case that it was driven far enough to the west, so that in the light of the development already made in the gangway next above, it was not likely to pay to drive it further. It probably went to the first of the series of little faults and rolls described above; for it stopped only about five hundred feet short of the Clark Vein main gangway itself.

To the east of the Pittsburg slope a similar state of affairs exists in the strike and dip to that above described in the western part of the Black Diamond company's mines, but on a somewhat smaller scale, and in an opposite direction. Though broken in the middle by a large fault, the gangways here run far to the south of east; and the dip also in the eastern part of the Pittsburg mine is considerably less than it is in the vicinity of the slope. Moreover, as already stated, the works at the end here abut directly against the wall of a great upthrow of some three or four hundred feet to the east.

VENTILATION.

In mines situated as these are, with a general dip of about 30°, among high hills and deep cañons, there is rarely much difficulty in securing good ventilation, if the matter be properly attended to; and the only artificial means in general use to aid the natural ventilation at the Mt. Diablo mines is the keeping of fires at the bottoms of the ventilating shafts. In only one instance has mechanical ventilation been resorted to on any considerable scale. In the Lower Mt. Hope gangway to the westward of a point about twenty-two hundred and fifty feet west of the Black Diamond shaft, all the water (and its quantity is considerable) which issues from the roof and floor of the Clark Vein, besides being a solution of sulphates, is supersaturated with sulphuretted hydrogen to such an extent that on exposure to the air it rapidly forms white deposits of sulphur everywhere; while the excess of gas escaping contaminates the air so much as to cause serious trouble by its effects upon the eyes, which it quickly renders sore, inflamed, and almost blind, probably by reason of its decomposition with the formation of minute quantities of sulphurous and sulphuric acids in contact with the moisture of the eyes. It was found impossible, with the ordinary means of ventilation employed here, to sent a sufficient volume of air through this portion of the mine to keep it clear enough of sulphuretted hydrogen to enable the men to work more than a very few hours at a time without becoming nearly blind. The trouble at last became so serious that the men absolutely refused to work there without better air.

The experiment was tried for a few days of keeping chloride of lime exposed to the air throughout that portion of the gangway affected, with a view to decompose and absorb the deleterious gas; but for some reason it appeared to fail to accomplish the work, while the men complained that the pungent odor of the chloride of lime, in addition to the sulphur gas, only made matters worse. Then one of the largest sizes of Root's patent rotary blowers was obtained, and, driven by a small steam engine, was set to forcing air through a pipe down the the Black Diamond shaft and into that part of the mine, the air afterwards finding its exit from the mine through a ventilating shaft connecting with the western part of the Mt. Hope Gangway. This made a decided improvement, but was yet far from being satisfactory. The action of the blower was therefore reversed, and it was made to exhaust the air, which then entered the mine through the ventilating shaft just mentioned. This did much better, and the work in this part of the mine was resumed and continued, though slowly and with difficulty; for the blower, after all, only partially removes the gas, enough of which still remains to make it very troublesome.

There is a little fire-damp in all the beds of the Mt. Diablo mines, and there are now and then localities which it is necessary to watch pretty closely. But the quantity of this gas has never been great enough to necessitate the general use of the safety-lamp in the workings; and it has, therefore, only been used as a precautionary means for testing the presence of the gas in certain localities where it was known that small quantities of it were liable to collect. Numerous small casualties, resulting in the more or less severe burning, and occasionally in the death of one or two men have, however, occurred here from time to time, from explosions of fire-damp, nearly all of which have originated in the gross carelessness of the sufferers themselves in going with naked lights into the top of blind shoots or other places which had been standing idle for awhile, and where they knew, or ought to have known, that fire-damp was liable to be present. But besides these minor casualties there have also occurred, during the history of the mines, three or four explosions of greater magnitude, which cannot be said to have resulted from the special ignorance or carelessness of individual miners. None of these explosions did much damage to the mines; but one or two of them have resulted in serious loss of life.

The worst one of all was the explosion, or more properly the burning, of July 24, 1876, on the Black Diamond bed, in the lower and eastern part of the Black Diamond Company's mines, which resulted in the death of eleven men. This was occasioned by a "blown out" shot. There was no explosive mixture present where this disaster occurred previous to the firing of the shot; for the men were working all along there with naked lights, and the ventilation was good and strong. But on the firing of this shot (which was a pretty heavy one, being a two and a quarter inch hole, charged probably with from twenty to twenty-four inches of black powder), the flame traveled between two and three hundred feet along the face of the coal, following a crooked course through cross-cuts, etc., developing a hardly noticeable amount of explosive force, but badly burning all the men whom it caught in its course, and then asphyxiating both them and others by the after-damp which followed the flash. The cause of this explosion was probably two-fold. First, it is well known that in coal seams containing fire-damp, any diminution in the atmospheric pressure, whether sudden or gradual, is accompanied by a correspondingly sudden or gradual liberation of increased quantities of fire-damp from the face of the coal. It is more than probable, therefore, that the recoil, which in an elastic medium like the air, and especially in confined localities, must instantly follow the first impulse of the heavy concussion of such a shot, would liberate suddenly from the adjacent face of the coal a certain quantity of fire-damp, which might issue forth quickly enough, and be sufficient in quantity to catch fire either from the flash of the shot itself, or from burning particles of coal-dust ignited by that flash. Second, the part of the mine where this accident happened was very dry, and the shot itself must have raised in its immediate vicinity a dense cloud of fine coal-dust. Now recent experiments have shown that a mixture of fire-damp and air which contains far too little fire-damp to be capable of either exploding or burning by itself alone, becomes readily explosive if mixed with a sufficient quantity of impalpably fine coal-dust. There is every reason, therefore, to believe that the propagation of the flame in this instance was effected by an intimate mixture of the air with a certain quantity of fire-damp and a dense cloud of coal-dust, the presence of the last two of which was mainly due to the concussion of the shot itself; and the whole affair is strongly illustrative of the danger of the use of powder in coal seams where fire-damp is known to exist.

The Mt. Diablo coal is liable, under favorable circumstances, to spontaneous combustion; and, in the Black Diamond bed, it is always necessary to shut up the old workings, so as to prevent access of air to the gob, which would otherwise heat and eventually take fire; but in the Clark bed, with its freedom from "bone" and its good sandstone roof and floor, no such precaution has been found necessary.

Haulage, Storage, and Transportation.

The gauge of the mine-tracks is different in the different mines. In the Pittsburg mine it is twenty-six inches; while in the Black Diamond mines it is three feet. The size and shape of the mine-cars employed also varies somewhat in the different mines. The inside dimensions of those used by the Black Diamond Company are six feet six inches long, two feet six inches wide, and two feet eight inches high. These cars are built of wood, banded with iron, and hold about a ton each of loose coal.

In the Black Diamond Company's mines, the underground hauling is done entirely by horses and mules; but in some of the other mines, and especially in the Union and Eureka, a great deal of it has been done by hand, the men pushing the cars.

No coal-breaking machinery has ever been used or needed here. In fact, one serious trouble with this soft coal is that it crumbles too easily and makes too much slack without any other breaking than that which it necessarily gets in mining and handling.

The bunkers at the mines are furnished with screens which separate the marketable coal into two sizes only, known as "coal" and "sceenings," respectively. But this screening is often very imperfectly done. The slack which falls through the finest screen is generally thrown away, though within the last few years considerable of it has been burned under the boilers at the mines, and occasionally a little of it has been sold for various purposes. The Black Diamond Company's bunkers are also furnished at the top with automatic dumping arrangements, so that the mine-cars dump themselves into the bunkers. The size of the bunkers varies, of course, with the requirements of the different mines. The largest one ever built here is that which receives the coal from the Black Diamond shaft. This bunker has a capacity of about sixteen hundred tons, *i. e.*, eleven hundred tons of coal and five hundred tons of sceenings. It stands at a distance of some five hundred feet or more from the mouth of the shaft, and the loaded cars are hauled to it and the empty ones hauled back to the shaft by an endless wire-rope worked by a clip-pulley, which is driven by a small steam engine. The floor of this bunker has a pitch of 33°, and the vertical height between the mine-car track above it and the railroad track beneath it is eighty-four feet.

The Black Diamond railroad is a trifle over five and eight-tenths miles in length. Of this, the first two and eight-tenths miles from the river to the edge of the hills is straight and has a grade which, though not uniform, being less near the river than it is near the hills, nevertheless averages for the whole two and eight-tenths miles about sixty feet to the mile. The remaining three miles in the cañon, from the edge of the hills up to Nortonville, is very crooked, and has an average grade of about one hundred and ninety feet to the mile. The maximum grade, however, is much heavier than this, and is situated at the upper end of the road nearest the mine, where, for a distance of five-eighths of a mile, the uniform grade is two hundred and ninety-three and three-fourths feet to the mile. The minimum radius of curvature in this road is three hundred twenty-one and one-fourth feet, corresponding to about an 18° curve. This, also, is at the upper end of the road, leading to one of the bunkers. Its use was necessitated by the position of the bunker and the narrowness of the cañon. It is just about as sharp a curve as the locomotives employed here will travel on without leaving the track.

The cars now employed on this road have flat wooden bottoms, with rectangular sheet-iron sides and ends strengthened with angle iron. One of the ends consists of a door, hung from a bolt which runs across the top, and furnished with a strong latch on each side. The cars are four-wheeled. Each car occupies about ten feet of track, and stands about six feet two inches high above the rails. The interior dimensions of the car body are as follows: length, eight feet; width, six feet four inches; height, three feet four and one-half inches. These cars weigh, on the average, about four thousand pounds apiece, and each car carries from ten thousand to eleven thousand pounds, or a trifle over four and one-half tons of coal. From twelve to sixteen of these cars form an ordinary coal train. The locomotive merely hauls the empty cars up to the mine. When loaded, the train runs down to the landing by its own gravity, and it needs, of course, careful attendance at the brakes to prevent it from running too fast. Indeed, whenever the track is muddy and slippery, as is often the case in the rainy season, it is found necessary, in addition to the most careful handling of the brakes, to sand the track to a greater or less extent before the descent of every train over that portion of the road which has the heaviest grade.

The Pittsburg railroad is of nearly the same length as the Black Diamond, and is very similar to it both in the distribution and in the amount of its grades and curvatures. The cars used upon this railroad are of iron, but of somewhat less capacity than those above described, and also somewhat different in construction, being arranged to dump through trap-doors in the bottom of the car, while the Black Diamond cars run on to a special dumping arrangement at the end of the wharf, which then tips up with the car upon it till the floor of the car makes an angle of 35° or so with the horizon, when the door at the lower end being unlatched, the coal runs out.

Pumping and Drainage.

There has never been any general system of drainage for the Mt. Diablo mines; but each company has pumped out its own water independently of all the rest. In the year 1869, I made a careful survey to ascertain what could be done in the way of a drain tunnel, and found that a tunnel about seven thousand feet in length from a point in the Somersville cañon to the Clark bed would drain all the mines of Nortonville and Somersville to a depth of only about three hundred feet above low water in the San Joaquin river, or in other words, to a point a little below the lowest levels ever yet reached by the workings in any of the mines, excepting those from the bottom of the Independent shaft.

This tunnel might have been driven for a cost of only about fifty thousand dollars, and at the time when the survey was made, there was talk of doing it. But the different companies to be benefited by it could not not agree as to the exact proportion of its cost which ought to be paid by each of them respectively; and so after talking awhile to no purpose, the matter was dropped; since which time the Black Diamond, the Union and the Pittsburg companies have all three of them been lifting their water to points over five hundred feet vertically higher than the level at which this tunnel would have drained them all; and it is safe to say that from then up to the present time, the aggregate cost of the item of pumping alone has been at least four or five times what the total cost of the tunnel would have been. And this is not, by any means, the only instance in which the existence and rivalry of so many different companies within so small a field, combined with short-sighted policy, and bad management in other ways, have caused the expenditure of large sums of money which are practically wasted, and which might otherwise have been saved to the owners of the mines. The Independent shaft was a bad job, as well as a bad speculation, from beginning to end; and the new Black Diamond shaft itself, with all its machinery, although a splendid piece of workmanship, was an unnecessary expense, inasmuch as the Mt. Hope counter-slope, which is large, commodious, well-timbered, protected by heavy pillars on either side, and furnished with a double track, and a pumping compartment besides, was already down to the same level as the present foot of the shaft, before the sinking of the shaft was begun; and it would have been easy to have hoisted through this slope, at no greater expense than it will now cost to hoist through the shaft all the coal that is likely to ever come up through the shaft.

It would be easy to point out many other ways in which money has been wastefully spent at the Mt. Diablo mines; but I will only mention here one other matter in this connection. Situated as these mines are, their whole extent from the eastern limit of the Pittsburg to the western limit of the Black Diamond company's workings, was none too great for a single colliery; and if in the early history of the mines the various companies had combined into a single organization to control and manage the whole, then not only the Black Diamond railroad itself, but also the whole establishment of shaft, slopes and machinery for pumping and hoisting at the village of Nortonville would have been needless and superfluous, and their entire cost might have been saved; for a single railroad in the Somersville cañon would have amply sufficed to transport all the coal which these mines have ever furnished, or ever will furnish; while, at the same time every ton of it could have been brought to daylight in this cañon more cheaply than it has been brought to the surface at the various openings through which it has actually been extracted.

For this purpose, the tunnel above referred to as a proposed drain tunnel, which was never driven, should not only have been driven, but should have been made a large sized working tunnel and furnished with a double track and all other facilities for the rapid extraction of coal.

I do not think it is too much to say that if this had been done, and if the general management of the mines had been at the same time placed in the hands of a competent engineer, then the total production of these mines up to the present time (which amounts approximately to one and three-quarter million tons) might have been furnished at an average cost price of one dollar less per ton than under existing circumstances its actual average cost has been. In other words, I believe that up to the present time the aggregate sum of one million seven hundred and fifty thousand dollars could have been saved to the owners of the Mt. Diablo coal mines, if all the natural advantages which the situation of the field presented had been utilized with the best economy.

Peacock and San Francisco Mines.

To the west of the Black Diamond Company's Mines, for a distance of a mile or two, there has been, in the past, considerable prospecting done for coal among the hills, and a number of slopes and tunnels have been driven some distance underground, at different points, upon the outcrops of small seams of coal. But nothing has ever been discovered here of any value, and the only two localities worth mentioning now are the old "Peacock" and "San Francisco" mines.

The first of these was unquestionably upon the Black Diamond bed, and is situated about a quarter of a mile south-westerly from the extreme western limit to which the Black Diamond Company pushed their old "Upper Black Diamond Gangway." This gangway, at its face, when abandoned, was running S. 46° 30′ W., true course, or about S. 30° W., magnetic, and the dip of the bed at the same point was about 20° to the northwest. It will also be remembered that the "Lower Black Diamond Gangway," as well as the corresponding levels on the Clark bed, in this portion of the Black Diamond Company's mines, all show the same great curve of the strata here to the south in going west. Yet there is no sudden break of any considerable magnitude so far as those works extend. But at the Peacock mine, only about a quarter of a mile distant, we find the bed striking S. 75° W., magnetic, and dipping to the north at an angle of 45°, thus proving that within this short distance there is a great disturbance of some kind, resulting in a sudden change of some 45° in the direction of the strike, and an increase of about 25° in the amount of dip.

I am not informed as to the full extent of the work which was done in the "Peacock," as it was already abandoned before I first saw it in 1868. But a slope was sunk for ventilation, and a tunnel was driven at least some eight hundred or nine hundred feet in length. The ground, however, was found to be badly broken and crushed, and the coal was soft and worthless.

The San Francisco mine is situated about half a mile to the west of the Peacock. A slope was sunk here about three hundred feet, on a bed whose dip is about $41°$ at the surface of the ground, but increases to $50°$ at the depth of one hundred and sixty feet, from which point a gangway was driven east two hundred and seventy-five feet, and west about seventy-five feet. At the eastern face of this gangway the coal was about two feet thick, and had a course of N. $70°$ E., magnetic, and a dip of $65°$ to the north. Below this stratum of coal there was about six feet of soft clay-rock, and then another stratum of coal about one foot thick. Above it there was also another small streak of coal separated from it by a layer of clay-slate. Small faults were very numerous here, and the coal was soft and friable, and never paid to mine. A little of it was once hauled to the village of Pacheco, for sale. But as early as 1869, the work was already abandoned. It is very probable, though not certain, that this mine also is on the Black Diamond bed.

Central Mine.

Passing now to the eastward from Somersville, the first mine which we encounter is the "Central" better known, perhaps, as "Stewart's Mine." This mine is situated in a steep and narrow ridge which was nearly east and west across the northern part of section 10, and the mine itself is in the north-east quarter of this section. It was originally opened by a level tunnel driven northerly from a point considerably beneath the line of outcrop of the beds in the steep and almost precipitous southern face of the ridge.

The course of this tunnel is just about north magnetic, and its length to the Clark bed is about one thousand and thirty feet. It of course cuts through the underlying beds. The tunnel is not at right angles to the gangway, the course of the latter being somewhat to the north of west magnetic. There are exposed in this tunnel, beneath the Clark bed, four distinct seams of coal, none of which, however, are here of any value. Starting from the Clark bed and going south along the tunnel towards its mouth, we find these seams as follows: The first one at a distance of one hundred and thirty feet shows eighteen inches of impure coal. The second one, about one hundred and fifty-seven feet further south, shows twenty-two inches of similar material. The third one, seventy-three feet further south, is nineteen inches thick; and the fourth one, supposed to be the Black Diamond bed, is one hundred and eighteen feet still further south. There can be no reasonable doubt, in spite of differences and diminished thickness in the section of the strata between them, that the two beds here called the Clark bed and the Black Diamond bed, respectively, are in reality the same beds as those which bear these names at Nortonville and Somersville. In this mine, some of the small beds between the two contain considerable gypsum in thin sheets and scales, filling seams in the soft and worthless coal. On the Black Diamond bed a gangway was once driven here some distance both east and west from the tunnel, and it is said that the bed was from three to four feet thick, and that about a thousand tons of coal were extracted from it, which, however, was of very poor quality, being both soft and "bony." It had been entirely abandoned previous to my first examination of the mine in 1869.

In April, 1870, a gangway had been driven here, on the Clark bed, two hundred and seventy-five feet east, and three hundred and seventy-five feet west from the tunnel, and a good deal of coal extracted, the bed averaging about thirty-nine inches in thickness, and dipping to the north at an angle of 28° or 29°. Just west of the tunnel, a large fault, consisting of an upthrow to the west, estimated at eighteen feet, crosses the gangway very obliquely, running north-west and south-east. To the west of this there were no more faults so far as the gangway was then driven, and the coal was bright and clean, but soft and friable. To the east from the tunnel there was a constant succession of small and irregular jumps all the way to the face of the gangway, and the coal here was badly crushed and very soft. Above this gangway the breasts had then been worked to a maximum height of about four hundred and fifty feet, the total distance on the dip of the bed, up to the outcrop, being about seven hundred feet. And some of the best, *i. e.*, the hardest, as well as the cleanest coal ever taken from the mine had come from the top of these breasts, up nearest to the surface of the ground.

In connection with the popular fancy that coal must of necessity improve indefinitely in quality with indefinite increase of depth beneath the surface, it may be well here to state the fact that at Nortonville, the mines have never, even from their lowest depths, produced any better or harder coal than was a great deal of that which came from the top of the very highest workings on the Black Diamond bed, more than five hundred feet above the old "Upper Black Diamond Gangway." And this is not all: It is true, as a general rule, throughout all the Mt. Diablo mines, that when a depth of from one hundred to three hundred feet is attained, measured on the dip of the bed from the outcrop, there is after and below that no further improvement in the quality of the coal which can be shown to be to any extent dependent upon or connected with the additional increase in the depth.

Since 1870, a tunnel has been driven in Stewart's mine from the Clark bed northerly entirely through the ridge and out to daylight on its northern side. Since the completion of this tunnel all the coal mined has been taken out through it, thus saving some two miles of cartage around and over the hill.

It is not probable that this mine has ever been a profitable one to work. And though it has produced in the aggregate a considerable quantity of coal, it has not been worked continuously, but irregularly and spasmodically, sometimes lying idle for many months, and then again producing as high as from nine hundred to one thousand tons of coal per month. After this sort of fitful life for some eight or ten years, it has recently again shut down, and it is doubtful whether it will ever be much more worked hereafter.

Going east from Stewart's mine, we next find in the

bottom of a cañon near Cochrane's house and close to the centre of section 11, the outcrops of two beds, which in all probability represent the Clark bed and the Black Diamond bed respectively. At this point the beds run very nearly east and west, and dip to the north at an angle which Cochrane states to be about 32°.

Some prospecting was done at this locality years ago, but the coal was not found good enough to warrant mining.

Beyond Cochrane's, as we go east, the thickness of the strata and the characteristics of the various beds themselves change so much that, though there is, of course, no lack of positive opinion on the subject among some of the men who are pretty familiar with the ground; and though there are here and there a few facts known which really do point to some probability in the matter with reference to certain beds, yet it is impossible, in the light of all the developments hitherto made, to recognize anywhere, with any certainty, a single bed as being identical with either the Clark or the Black Diamond bed of the Mt. Diablo mines.

The next development to the east of Cochrane's, is in the north-east part of the south-west quarter of section 12. Here a slope was sunk about two hundred feet some years ago, in a direction of north 16° west magnetic upon a bed of coal with a pitch of about 27°. There was no coal visible here at the surface of the ground, but only a slightly carbonaceous shale for the first eighty or ninety feet. But then the coal began to come in, and at the bottom of the slope there is said to have been three feet of pretty clean, though rather soft, coal, with a good sandstone roof. It is also said that two small schooner loads were once shipped from the bottom of this slope.

Empire Mine.

The next development is at the locality now known as the "Empire Mine." This is in the south-west part of the south-east quarter of section 12.

A slope was originally sunk here about two hundred feet in 1860 or 1861, when the work was stopped by the influx of water which the parties had not the means to handle. There was visible here at the surface of the ground only a little streak of soft clay-shale about eight or ten inches thick, which was of rather a light yellowish hue, being but very slightly colored by carbonaceous matter, and having sandstone immediately above and below it. This could not be called a very promising outcrop, certainly. But, on going down, it was found that this streak of shale increased steadily and rapidly in thickness, and also grew rapidly more and more carbonaceous, till, at the depth of one hundred feet slope distance, it had already developed into a four and a half foot bed of what might very properly be called coal, though it was still impure and very soft and friable. Its quality still continued to improve rapidly to the bottom of the slope. It was, however, abandoned.

But in the year 1875, Mr. George Hawkshurst, the superintendent of the Union mine, at Somersville, in connection with one or two other parties, again took hold of this old slope, cleaned it out, enlarged it, furnished it with a double track, put up pumping and hoisting machinery, and sunk it to the depth of six hundred feet (slope distance), and then drove a gangway both ways from its foot.

My last visit to this property was December 11, 1876. At this time the gangway was driven about three hundred feet west and nearly four hundred feet east from the slope, with a general course of N. 75° E., magnetic, the dip of the bed being about 23°, and the direction of the slope itself being N. 6° E., magnetic.

The coal along this gangway ranges from three feet six inches to a little over four feet in thickness. At the west face of the gangway it was four feet three inches thick. Of this, the upper twelve inches was tolerably clean coal; the next twelve inches was "bony," and the lower two feet three inches was clean coal, though rather softer than the average Mt. Diablo coal. The floor of the bed is sandstone. Along the roof of it there runs a stratum of from five to eight inches of soft clay-slate, which, however, is not continuous, the solid sandstone sometimes coming down to the coal. Above this little streak of slate there is everywhere good solid sandstone. In the eastern part of the gangway there is one fault, which consists of a down-throw to the east of just about the thickness of the vein. West of the slope, there are only one or two little jumps, of a few inches each.

From a point a few feet east of the foot of the slope, a tunnel was driven south some three hundred feet through the sandstone, in order to strike an underlying bed which had been previously discovered by a little shaft sunk about ninety feet south of the mouth of the slope and one hundred feet deep. This bed, as seen in the shaft, is said to consist of three feet of good clean coal, like the bottom bench of the upper bed, without any "bone" and with good sandstone roof and floor. This bed they had not yet reached in the tunnel at the time of my visit, though at a distance of a little less than two hundred feet from the upper bed they had passed through a small coal seam about eighteen inches thick. Since that time, however, they have struck the lower bed in the tunnel, and found it, as I am told, to consist here of a bottom bench of twenty-two or twenty-three inches of clean coal, overlaid by about fourteen inches of worthless "bone." The appearance of this "bone" at the depth where the tunnel strikes it, while there was no "bone" at the bottom of the little shaft so much nearer the surface of the ground, is not an encouraging fact with regard to the future prospects for a mine upon this bed.

At a point some six hundred feet south of the mouth of the slope and very close to the section line between sections 12 and 13, there has been another little shaft sunk about ninety feet, and from the bottom of it a drill-hole was pushed some thirty feet lower. They are reported to have passed through several little streaks of coal in this shaft, and at the bottom of the drill-hole to have struck something which they believe to be the Black Diamond bed, as they assume the bed upon which the slope is sunk to be the Clark bed, and the one struck in the tunnel to be one of the "Little Veins" between the two. But this assumption, though not improbable, is, as already stated, by no means proven.

A recent survey shows that the mouth of the Empire mine is about four hundred feet above tide-water, and that a railroad from there to the village of Antioch, on the San Joaquin river, will be about five and a half miles long, and will have two tunnels, aggregating something over one thousand feet in length. It is the present intention of the owners to build this road.

TEUTONIA MINE.

Next east of the Empire mine comes the old "Teutonia." This is in the south part of the south-west quarter of section 7 of township 1 north, range 2 east, the mouth of the mine being only about one hundred and fifty feet north of the section line. This mine was furnished with steam hoisting and pumping machinery. But at the time of my first visit to it in September, 1869, it had already been idle and abandoned for some two years, and nothing has been done there since. According to the best information which I have been able to obtain, however, relating to this mine, the slope, which was furnished with a double track and with sheet-iron mine-cars, went down upon a bed of coal about four hundred feet, with a pitch of about $26°$. From the bottom of the slope a gangway was driven east something like one hundred feet. Just west of the slope the bed was broken by a large fault jumping up to west, beyond which the work was never carried. The bed was about thirty-six inches thick, the lower half of it being bright, clean, shelly coal, not very hard, and the upper half being "bony." It will be noticed that this description of the bed itself is remarkably like that of the bed which was struck by the tunnel in the Empire mine in the latter part of December, 1876; and it is indeed not at all unlikely that it may be in reality the same bed.

The fact is worth noticing here that on October 11,

1875, before the underlying bed had been found at the Empire mine, Mr. J. Cruikshank (who is well informed as to the early work which was done in this region), in some notes which he gave me, placed the Teutonia slope upon a bed underlying the "Clark Vein," and located the outcrop of the "Clark Vein" itself at a point some distance to the north of the mouth of the Teutonia slope.

On the north-east quarter of section 18, township 1 north, range 2 east, there is another old slope, known as the "Israel Opening." This slope is said to be some two hundred feet deep, with a pitch of about 25°. It is said, furthermore, that at its bottom there was three feet of clean and tolerably hard coal, and that some rooms were opened and several cargoes of coal once shipped from here. It is supposed that this slope is on a bed which underlies the one on which the Teutonia slope is sunk.

On the north-west quarter of section 16, township 1 north, range 2 east, there are several small openings, only one of which is worth mentioning now. This is a slope which runs down about north magnetic with an average pitch of 21°. It is said to be about two hundred feet deep, and also that at the bottom of it there were three feet of clean coal, with sandstone roof and floor. In December, 1876, the lower part of this slope was full of water, down to the surface of which it was one hundred and thirty-five feet, and at this depth there was nothing like good coal to be seen, but only a streak of dirty "croppings," about one foot in thickness.

Rancho de Los Meganos.

On going still further to the east from here, there is for some distance hardly any exposure of the rocks at the surface, and there have never been any holes sunk until we reach the south-east quarter of section 22, and the north-east quarter of section 27, upon the Rancho de Los Meganos, in township 1 north, range 2 east. Here there are known to exist at least three beds of coal of workable thickness associated with heavy deposits of a good quality of fire-clay.

A small shaft in the south part of section 22, known as the "Hoisting Shaft," and eighty-eight feet in depth, shows the following section of the strata, the measurements being vertical, and beginning at the top or mouth of the shaft:

	Feet.	Inches.
Clay and clayey material	34	4
Black clay	14	8
Coal	2	4
White clay, hard and somewhat sandy	4	8
Coal	0	4
Blue fire-clay	5	0
Coal	3	6
Clay (with three regular coal-seams, about one foot thick each)	8	0
Coal	7	0
Clay	3	0
Coal	1	2
Clay	4	0

There has been mined here, chiefly from the "7 foot" and the "3½ foot" beds, through shallow slopes and shafts, without the use of other power than that of hand and horse, an aggregate of probably somewhere between five thousand and ten thousand tons of coal, most of which has been used under the boilers at the "Engine Shaft."

The general course of strike of the beds here is about N. 72° W. magnetic, and their dip to the north-east, but so far as yet explored somewhat variable in amount, ranging from 16° to 26° at different points.

The "Engine Shaft" is sunk at a point about eleven hundred feet north-easterly from the line of outcrop of the beds, is about three hundred and eighty feet deep, and is divided into three compartments, two hoisting and one pumping, each compartment being 8 ft. × 5 ft. clear inside of timbers. The shaft is well timbered and is a good piece of workmanship. At its bottom there is a seven-foot bed of coal upon which a gangway was driven west in 1868, to a distance of two hundred and seventy-five feet from the shaft. No gangway was ever driven east from the shaft, and the foot of the shaft itself is in a fault which appears to be an upthrow to the east, of considerable magnitude. Very little coal was ever mined from here, and what was taken out was burned under the boilers at the shaft. The quantity of water to handle here was pretty large, and the shaft was furnished with a Cornish pump, the pumping engine having a 22-inch cylinder with 48-inch stroke, and being geared 4 to 1. The hoisting engine has a 16" × 48" cylinder and is geared 3 to 1.

It was but a few months after reaching the coal at the foot of this shaft, when, the company which owned the property getting into financial trouble, the work was suspended, and the shaft allowed to fill with water.

Since that time it has been once again pumped out, and kept clear of water for a month or two, when, owing to similar causes, it was again allowed to refill. And in this condition it has remained up to the present time, the water standing about forty feet below the mouth of the shaft.

It is believed by Mr. R. F. Lord, the engineer in charge of this property since 1871, as well as by Mr. Clarence King, mining geologist, who made a report upon it in 1874, to Mr. S. E. Lyon of New York, that the seven-foot bed at the foot of the engine shaft is entirely a different and separate bed from any of those upon which any mining has been done in the shallow workings near the outcrop, and that the latter beds, denominated by King the "Lord Series," underlie the former, the vertical thickness of the strata between the upper and lower seven-foot beds being supposed to be about one hundred and twenty-five feet.

But while this theory is not *a priori* particularly improbable, it is yet far from being proven to be true, and it is based upon facts which, after a recent careful examination of the ground by myself, and with my experience of over nine years of intimate acquaintance with the coal mines of the Mt. Diablo region, I consider to be of very questionable import, and of little value.

It would be nothing wonderful if this seven-foot bed at the foot of the engine shaft (which bed consists, by the way, of three distinct benches of coal, separated from each other by two layers of clay-slate a few inches each in thickness), should eventually turn out to be identical with, and at this depth the only representative of, the whole series of beds which has been called the "Lord Series." But it is a question upon which the paucity and the doubtful significance of existing developments render speculation idle, and which additional underground explorations alone can finally settle. Whatever the fact may prove to be, however, in this respect, there can be no question in any case that the quantity of coal in the Rancho de Los Meganos is great. And, though I have never seen any coal in this property which was quite so hard or which would bear handling and transportation so well as the average of the Mt. Diablo coal, nevertheless, as it can be cheaply mined and cheaply sold, there is good reason to believe that it will pay to open up and work this mine, so soon as the property shall be freed from legal complications and a clean title shall be vested in some party who has both the money and the intelligence which it will certainly require to handle it properly.

With the Rancho de Los Meganos, the Mt. Diablo coal field may be said to terminate, no explorations to the east or south-east of here having ever developed anything in the shape of coal worth mentioning until we come to another field, viz.,

THE CORRAL HOLLOW COAL FIELD,

In the hills to the south of the Livermore Pass. There is a general description of this coal field, together with the developments which had been made here in the way of exploring and mining for coal up to the year 1862, in the volume of *The Geology of California*, published by the State Geological Survey in 1865, pages 34 to 38, to which the reader is also referred.

In the year 1870, I visited the locality myself, in the employment of the State Geological Survey. In the eight years which had then intervened since Mr. Brewer's last visit, there had been considerable work done and a good deal of money expended in prospecting and mining for coal in the Corral Hollow cañon, the results of which had only tended, however, to confirm the accuracy of the opinion expressed on page 36 of *The Geology of California*, that "the disturbances of the strata in this district were so extensive that it was to be feared that these coal beds would not be made available;" while the quality of the coal itself had also been proven to be somewhat inferior to that of the Mt. Diablo mines, inasmuch as it is softer and more friable, and crumbles worse upon exposure to the atmosphere.

At the old Pacific mine (otherwise called the "Eureka mine," and "O'Brien's mine,") no work had been done since 1862. Farther down the cañon, though my notes of the trip show many detailed observations of the strike and dip of the strata, as well as of the other visible surface indications, the only mention it is worth while to make of them here is the fact that they all confirm the statement that the strata are greatly disturbed. All the lower mines were already, at the time of my visit, in August, 1870, entirely closed and abandoned; and the best information I could obtain respecting the underground developments in them was from Mr. Carroll, who had lived here for some years, and was pretty familiar with the work that had actually been done. According to his statements, the old shaft of the "Commercial Company" (which is situated on the south side of the creek, some half or three quarters of a mile below the shaft of the old "Coast Range Company," described on pages 37 and 38 of *The Geology of California*), was sunk about two hundred feet, and a tunnel was driven from its bottom about one hundred and eighty feet to the south. This shaft was not in coal, and the tunnel from its foot did not strike coal. A short distance below this point there is another shaft sunk to a depth of about eighty feet by Mr. Meader. This also was not on coal, and no drifting was done from it.

The next opening which we come to is the "lower shaft" of the Commercial Company. This shaft is in the coal, is about three hundred feet deep, and furnished all the coal which came from the Corral Hollow mines during the years 1869 and 1870. But at the time of my visit the hoisting-works had recently been burned down, and the mine itself, as well as its waste heaps, was on fire. The dip in this mine was very steep to the south.

At the Almaden mine, a little further down the cañon, there is a shaft about three hundred feet deep, and a tunnel was driven southerly from its foot about seven hundred feet, but no coal was found, except two or three seams of no value. Carroll thinks there is coal here to the north of the shaft. The dip here is southerly.

While this work was going on previous to 1870, the Western Pacific Railroad Company had also expended a few thousand dollars in laying down a track from Ellis Station to the mouth of Corral Hollow cañon, in the hope of getting coal from these mines for use upon their locomotives, in which hope they were, not unnaturally, disappointed.

I have not heard of any further mining for coal in Corral Hollow cañon since 1870; and the total amount of coal ever sent to market from this locality has been very small.

But outside of Corral Hollow cañon, and yet within the limits of what may be properly called the Corral Hollow coal field, there has been some prospecting and a little mining done.

The Livermore Mine.

Within about a mile to the west of the Pacific mine, and on the west side of the crest of the watershed which here divides the waters of the Livermore valley from those of the San Joaquin valley, there is situated the "Livermore Mine." At this mine, when I visited it in July, 1875, they had sunk a slope of about three hundred and eighty feet upon a bed of coal whose strike was just about east and west magnetic, and whose dip, though somewhat variable, averaged to the bottom of the slope about 40° to the north.

At the surface of the ground, there was visible here only a little black dirt; but the coal began to come in at a point about fifty feet below the mouth of the slope. At the bottom of the slope, when I saw it, the bed was about five feet thick, but contained three or four little streaks of clay, from half an inch to three inches thick. The coal itself was soft, and crumbled on exposure, like that of the mines of Corral Hollow cañon. Some further work was done here in the latter part of 1875. A steam hoisting engine was erected, and bunkers were built, and some drifting was done underground, but the work has since been abandoned.

It is reported that since 1875 another coal discovery has been made at the so-called "Summit Coal Mine," a short distance to the north-east from the Livermore mine, and that considerable prospecting work has been done there, with promising results. But of this I cannot speak positively, not having seen the ground.

OTHER COAL LOCALITIES.

Outside of the Mt. Diablo coal field, there are numerous localities besides Corral Hollow scattered throughout the coast range of mountains from San Diego to Crescent City, and a number of localities also in the western foothills of the Sierra Nevada, in California, where more or less coal has been found. None of these localities have yet proven themselves to be of any financial value here, and the great majority of them would be utterly worthless in any country. I proceed, however, to mention a few, which either from their own intrinsic merit, or else from the noise which has been made about them, are worthy of special notice.

First. In the southern part of Los Angeles County, at a locality about twelve or thirteen miles easterly from the town of Anaheim, in the mountains on the south side of the Santa Ana River, not over a mile from the river, and at an altitude of some fourteen hundred or fifteen hundred feet above its bed, there are exposed in the precipitous mountain-side some ten or twelve thin seams of impure coal, distributed through something like a hundred feet in thickness of shales and sandstones, no single coal seam being over about one foot thick. I visited this locality in 1872. The whole thing is worthless.

Second. It is said that there is a locality upon Los Gatos Creek, in the eastern flank of the Coast Range, in the southern part of Fresno County, where there are exposed no less than four or five beds which show in their croppings from three feet to four and one-half feet respectively of a good quality of coal, which it would pay well to mine if it were within reasonable distance of a market. This locality I have not seen.

Third. There is in the hills on the south side of the little valley called Vallecitos, in the western part of Fresno County, and distant some five or six miles in a north-westerly direction from the New Idria Quicksilver mine, a bed of coal which strikes N. 85° W., magnetic, and dips 80° to 85° to the south. This bed is certainly over seven feet in thickness, as, at the time of my visit in April, 1871, it had already been pierced to that extent by a tunnel which had not yet gone through it. This tunnel struck the bed at a depth only about forty feet below the surface of the ground. So far as exposed at that time, the coal was pretty uniform in quality throughout, and appeared but little contaminated with earthy matter. It however contained considerable gypsum in thin scales filling its seams, and it was soft and friable. But its quality was good enough, on the whole, to warrant a belief that it might, with proper arrangements, be used to some extent with advantage in the reduction of quicksilver ores at the New Idria mine, where wood is scarce and expensive, though whether since then it has actually been so utilized, I am not informed.

Fourth. On the Middle Fork of the Eel River, about seven or eight miles south of the village of Round Valley, in Mendocino county, and in the north-east corner of section 11 of township 21 north, range 13 west, Mt. Diablo meridain, there is a bed of coal exposed, crossing the channel of the river in a direction N. 45° W. to N. 50° W. magnetic, and dipping from 20° to 30° north-east.

This bed is from fourteen to fifteen feet thick, and is all good coal with the exception of a single streak of shale in the middle of it, about five or six inches in thickness. The coal is immediately overlaid and underlaid by heavy beds of very fragile shales.

The shales above the coal are not far from seventy-five feet thick, and are overlaid by very hard and highly metamorphic rocks, containing large quantities of jasper and other silicious matter.

The shales beneath the coal are about twenty feet thick, and are underlaid by a bed of unaltered sandstone some ten or twelve thick, which again rests upon the same hard, metamorphic rocks which overlie the shales above. The whole thickness, coal and all, therefore, of the belt of unaltered strata which includes this coal bed, is at this locality only about one hundred and twenty-five feet.

The quality of the coal itself is a little better than that of the Mt. Diablo mines. In fact, it is the best coal which I have seen from anywhere in California; while at the same time this is the thickest bed of a marketable quality of coal that is yet known to exist within the state. Two causes, however, combine to render it improbable that it will ever furnish coal for the San Francisco market. In the first place, there is plenty of evidence close at hand that the rocks in that neighborhood have been greatly disturbed, and it is very uncertain how far the bed could be followed without being found crushed and broken up by faults; while at the same time extensive metamorphism of the rocks has been peculiarly localized and capriciously distributed throughout this region, and very irregular patches and belts of highly metamorphosed rocks alternate in all directions with no less irregular belts and patches which seem to have almost entirely escaped the metamorphic action. And, in the second place, the locality is in the heart of the Coast Range of mountains, and in order to reach it, it would be necessary to construct a railroad for a long distance through a very rough region, which would render the cost of transportation so great that coal can be laid down in San Francisco from Washington territory or Vancouver's Island for less cost per ton than from here. There are said to be several other localities to the west and south-west from Round Valley where some croppings of coal have been found, but none of these are of any special interest.

Fifth. In the eastern part of Shasta County, there is among the western foothills of the Sierra Nevada a region of considerable extent, including portions of several townships, where the volcanic materials which cap the mountain spurs and ridges are generally underlaid by a body of coal-bearing strata of recent origin. These strata consist of soft and unaltered shales and sandstones, and they are spread out unconformably over the upturned edges of the metamorphic gold-bearing slates which form so large a part of the mass of the Sierra. Their general position is not far from horizontal, though at different points they dip gently in various directions, the angle of dip rarely, if ever, exceeding $6°$ or $8°$. The aggregate thickness of these strata is probably not over one or two hundred feet; and they belong to that geological period which immediately preceded the commencement of volcanic activity in that portion of the range.

At a point in the north-west quarter of section 20, township 33 north, range 1 west, Mt. Diablo meridian, there was in September, 1874, an open cut in a hill side, thirty-five feet long, beyond which a tunnel had been driven fifteen feet underground; and in this tunnel there was exposed a coal-bed whose total thickness was twelve feet. This thickness was made up as follows, beginning at the top:

	Feet.	Inches.
Coal, slaty and worthless	1	6
Slate	0	6
Coal	0	7
Slate	0	5
Coal	0	11
Slate	0	3
Coal	1	2
Slate	0	3
Coal	0	1
Slate	1	10
Coal	0	5
Slate	0	4
Coal	0	4
Slate	0	2
Coal	1	10
Slate	0	4
Coal	0	4
Slate	0	3
Coal	0	6
Total	12	0

What is here designated as "coal," however, was itself more or less impure, being often traversed by still thinner sheets of clay and dirt, whose thickness ranged from that of a sheet of paper up to half an inch, or so. It was also soft and friable, and disintegrated rapidly on exposure to the air. This to be sure was very close to the surface of the ground.

When I again visited the same locality, in April, 1876, this tunnel had been driven some thirty feet further underground, and then allowed to cave, and the place was inaccessible. I was told by Mr. Kincaid, who did the work, that at the face where he last stopped the coal was somewhat harder, and contained less slate than where I saw it in 1874. But heavy as this bed is, its quality at the best, so far as yet explored, is such that unless it improves very materially on driving further into the hill, it is not likely to pay to mine.

In the near vicinity of this point, also, there has been considerable other prospecting work done, and one tunnel has been driven some four or five hundred feet in length. But none of this work has developed so much coal as the open cut and tunnel just described.

I also saw more or less of coal croppings at various other localities scattered about through this region. For example, on section 3, section 7, section 8, and section 21 of this same township, also on section 12, township 33 north, range 2 west, and also at a point which is probably in section 9, township 34 north, range 1 west. Croppings are said to be exposed also in section 27 and section 28 of township 33 north, range 1 west. But very little work has been done, however, at any of these localities, and no coal has yet been found which it would pay to mine.

Sixth. In Ione Valley, at the western edge of the foothills of the Sierra Nevada, in Amador County, there is a coal-bed, which has attracted some attention, at a locality which I visited incidentally in November, 1871, while more especially engaged in studying for the State Geological Survey the ancient auriferous gravels, which are so widely distributed over the western flanks of the Sierra.

This coal is also of very recent origin; quite probably, indeed, not older than some of the auriferous gravels themselves. The bed lies nearly horizontal, and ranges at different points from five or six to twelve or fifteen feet in thickness. It is overlaid and underlaid by a very soft clay-rock, and its depth beneath the surface of the ground is small, being sometimes not more than thirty or forty feet. The material itself is strictly a lignite, still showing a good deal of the woody texture. It is not black nor lustrous, but of a dull earthy brown color, very soft and friable, and makes a large quantity of ash. Nevertheless, it burns very freely with a bright flame, and the ashes do not form any troublesome clinker. It has been employed for years as fuel for a flouring-mill at Ione City, the distance to haul it being about three-quarters of a mile. At the time of my visit, this mill, driven by a steam engine of 14″ cylinder and 36″ stroke, was using no other fuel, and was consuming of this, as Mr. Hall, the proprietor, informed me, about three tons per day, costing less than a dollar and a half per ton at the mill. This was certainly very cheap fuel; and the Ione Valley coal will be likely to continue for many years to supply a certain moderate local demand for various purposes; but it will not bear transportation to any great distance, and it is not likely to ever compete with other coals in the general market. Since the beginning of 1876, a new mine has been opened here, and there has been a good deal of talk about it; but whether the quality of its coal is in reality any better or poorer than was obtained from the earlier workings I do not know, not having yet seen the mine myself.

Seventh. At the village of Lincoln in the Sacramento valley, in the south-western part of Placer county, there is also a coal deposit, of which great expectations have from time to time been entertained. I have never examined this deposit and do not know the extent of the work which has been done. But I have seen some of the coal which it has furnished and such of it as I have seen was decidedly inferior in quality even to the Ione Valley coal; so poor, in fact, as to be practically worthless.

Eighth. At American Cañon, in the south-western part of Solano County, there are, for some distance in the bluff along the right bank of the cañon, heavy but irregular croppings of black carbonaceous shale, containing streaks from one inch to eight or ten inches in thickness of coal. Most of these croppings, however, are not in place, as there has been more or less land-sliding nearly all the way along the steep face of the bluff.

The attempt has been made once or twice to organize a company to mine here for coal. But there has never yet been sufficient work done here to prove what lies in the solid hill back of the croppings. The locality would also be rather an expensive one to prospect satisfactorily, and the surface indications are not on the whole particularly promising. With reference, however, to transportation and proximity to market, the situation is a very favorable one if ever a good mine be found here.

Ninth. There have been occasional paragraphs in the newspapers, within the last year or two, with reference to the discovery of what has been asserted to be a heavy bed of a superior quality of coal in the range of hills next east of the Santa Rosa Valley in Sonoma County. But I am not aware that this discovery has yet proven itself to be of any value.

Tenth. In addition to all the foregoing, there have been numberless "coal discoveries" reported in the newspapers from time to time, in almost every corner of the State; but more especially in the Coast Range of mountains, and more particularly still in the counties of Santa Cruz and Monterey, and in the Contra Costa hills which stretch south-easterly from Carquinez straits through Contra Costa and Alameda counties, and in the foothills which skirt the southern and western flanks of Mt. Diablo itself. And in very many, probably indeed in nearly all of these numerous localities, a little coal of some sort has actually been found. But none of them all have yet proven to be of any practical value, and the statement still remains true to-day, as it has done in the past, that the only locality in California where coal has ever yet been mined with profit to any noteworthy extent, is at the old Mt. Diablo mines.

But it is furthermore true to-day, of the Mt. Diablo mines themselves, that all of them which have been profitable in the past have already seen their best days and are now rapidly declining; while outside of these old mines the most promising region yet known in the State is the eastern and yet unworked part of the Mt. Diablo coal field, in which the most promising developments yet made are at the Empire mine and at the Rancho de los Meganos.

CHAPTER II.

OREGON.

THE COOS BAY MINES.

The only coal mines hitherto worked with profit in Oregon are located at Coos Bay, on the western coast of the State, about one hundred miles north of the California boundary, and some forty miles to the north of Cape Blanco.

The full-extent of the Coos Bay coal field is not yet definitely known, and it will probably be many years before it will become so. The wild nature of the country, which is everywhere covered with dense and heavy forests, renders exploration extremely slow, laborious and expensive; while the distance from market, combined with the difficulties of navigation make it impossible to mine coal here with profit, except at a few points, where it is found of good quality, favorably situated for cheap mining, and very close to navigable water.

It is already known, however, that this coal field covers at least several hundred square miles of territory, stretching from the mouth of the Umpqua river on the north to points beyond the Coquille river on the south, and extending back from the coast to distances of from fifteen to twenty miles inland.

The general character of the surface of the country thus designated is that of a hilly region, carved into a perfect labyrinth of steep and narrow gulches, and in the vicinity of Coos Bay, pierced in every direction by numerous "sloughs," or long narrow and crooked inlets, which though usually not very deep, often allow the tide-water to set back in them many miles. The lower portions of the Umpqua and the Coquille rivers themselves also assume the character of these so-called "sloughs," and the tide ebbs and flows in them for a considerable distance back from the coast.

Over the greater portion of this region the hills rarely rise to altitudes of more than five or six hundred feet above tide-water; but the banks of the sloughs are generally abrupt and steep, and the gulches which cut the hills in every direction are so numerous as well as deep and steep, that with the heavy timber and frequently the dense underbrush besides, the country is a very hard one to prospect, or to travel through.

As a general rule, along this portion of the coast, each river or inlet has in front of it a bar of sand, upon which the water is not so deep as it is in the sloughs inside. These bars, moreover, are shifting, and the deepest channels over them vary more or less in position as well as in depth, with every heavy storm. The depth at low water on the bar in front of Coos Bay varies, in different seasons, from nine to thirteen or fourteen feet; and it is a most serious disadvantage to the Coos Bay coal mines that this depth is so small as not to permit the general use of vessels for transporting the coal, which will carry more than from three hundred to four hundred tons at a cargo. Moreover, in the winter the sea is frequently so rough that it is unsafe for vessels of any kind to attempt to cross the bar; and it has happened more than once that vessels laden with coal and lumber have been land-locked in Coos Bay, and unable to get out for over a month at a time.

Coos Bay itself is roughly crescent shaped, convex towards the north. From the head of the bay proper, which is about at the village of Marshfield, Isthmus Slough continues navigable some six or eight miles farther south for vessels as large as can ordinarily cross the bar. There is thus included between the upper and lower portions of the bay, and between Isthmus Slough and the ocean coast a peninsula, which in its southern portion is eight or nine miles wide. Empire City, on the lower portion of Coos Bay, is on the north-west side of this peninsula; the village of North Bend is at its extreme northern point; while Marshfield is on its eastern side, and, as already stated, at about the head of the bay itself.

There is a saw-mill at Empire City, another at North Bend, and a third at Marshfield. The united capacity of these three mills amounts to not far from fifty thousand feet of lumber per day. There has been some shipbuilding done here; and North Bend has turned out one or two remarkably fast sailers. A large proportion of the so-called Port Orford cedar, viz., the white cedar of Oregon, which reaches the San Francisco market, comes from the Coos Bay mills. All three of the villages named above are in township 25 south, range 13 west, Willamette meridian. The paying coal mines of Coos Bay are all in township 26 south, range 13 west. There are only two mines which have been profitably worked in the past, viz., the Eastport mine and the Newport mine, and there is at the present time but one more which gives any reasonable promise of profit in the future.

THE EASTPORT MINE.

At a point on the left bank of Isthmus Slough near its mouth and a mile or so above Marshfield is the mouth of "Coal-Bank Slough." From here up the latter slough, which is small and crooked and navigable only at high water, it is about three-fourths of a mile to the point of shipment of the coal from the Eastport mine. This mine is situated in the north-east quarter of section 4 of township 26 south, range 13 west, though it is opened by a tunnel whose mouth is in the north-west corner of section 3; while its most northerly underground workings probably also extend somewhat beyond the township line on the north into the south-east quarter of section 33 in the adjoining township. The height of the mouth of the mine above tide water is stated to be one hundred and sixty or one hundred and seventy feet, and the length of the railroad from there to the landing on Coal-Bank Slough is about seven-eighths of a mile. This railroad is a wooden tramway of about four feet gauge furnished with a strap iron rail. The cars are of wood, with trap door discharges at the bottom, and hold about two and a half tons each. Each empty car is hauled separately from the landing up to the mine by a horse or mule. The loaded cars run down to the landing by their own gravity, and four of them being coupled together constitute a train, the brakes of which are managed by one man.

The coal bed here strikes about true north and south, and throughout the upper portions of the mine has an average dip of about 8° to the west. The aggregate thickness of the coal mined ranges from four to five feet, averaging about four feet and a half, in two benches of nearly equal thickness with a stratum of soft clay-slate between them whose average thickness is about six inches, though it ranges in different parts of the mine from four to ten. Above the upper one of these two benches comes a stratum of clay-slate, generally about one foot in thickness, and immediately over this again about one foot of coal. But this so-called "top coal" is not generally taken down, for the reason that the material overlying it is a soft and weak clay-rock which would not form so good a roof for the mine as the slate beneath the top coal does, while at the same time that coal, though generally of good quality, is not thick enough to pay for handling so much rock as would be necessary in order to obtain it.

This mine is opened by a tunnel driven westerly from a point below the outcrop of the bed a distance of some seven hundred feet to where it strikes the coal. From this point a gangway has been driven a considerable distance both north and south, and all the available coal above its level taken out. From a point in this gangway some five or six hundred feet south of the tunnel a slope has been sunk following the dip of the bed to the west a distance of about nine hundred feet. From this slope gangways have been driven north and south at various points, some of them to distances of two thousand feet or more, and the upper lifts thus obtained are also by this time pretty well exhausted. At the time of my last visit to this mine, in May, 1876, the gangway from the foot, or extreme western end of this slope, had extended about two hundred and fifty feet north and three hundred and fifty feet south from the slope. In going down this slope the dip of the bed itself gradually decreases from about 8° at its head until within about one hundred feet of its foot it vanishes altogether, and for the remaining distance the coal lies just about level. At the same time, however, the bed becomes somewhat more irregular and shows numerous small rolls with occasional faults, which make it more expensive to mine, while the coal itself is not quite so hard and good as it was in the upper levels of the mine.

The Newport Mine.

The Newport Mine is in the northern part of section 9 of township 26 south, range 13 west, and its mouth is only about ninety feet above tide water. Its point of shipment is also on Coal Bank slough, at some little distance above that of the Eastport mine. The railroad from the mine to the landing is a little less than two miles in length, laid with the strap rail, with a gauge of three feet two inches, and for a portion of the distance it has an up grade for the load of about one hundred feet to the mile in order to gain sufficient height at the landing for shoots to load the vessels. It is furnished also with a little four-wheeled locomotive, only two of the wheels being drivers, which is said to weigh seven tons and to have cost about four thousand dollars. This engine hauls four cars at a time, containing ten tons of coal, from the mine to the landing.

The Newport Mine is opened by a tunnel or gangway, which starts in the croppings of the bed on the south side of a ravine and runs directly in the coal itself for a distance of over two thousand feet to the south, entirely through the spur of the hills in which the mine is located and out to daylight on the other side. No coal has yet been mined here below the level of this gangway. The maximum height of the lift above it in the middle of the spur was probably over fifteen hundred feet. But the greater portion of this is now worked out, and in May, 1876, the foreman of the mine informed me that the quantity of available coal then remaining above this gangway would not exceed a two years' supply at the rate of fifty tons per day. Another gangway can be driven here, however, at a lower level which will give an additional lift of two or three hundred feet below the present one and thus furnish several years' further supply of coal before it will be necessary to do any pumping or hoisting.

The bed which is here worked is unquestionably the same as that in the Eastport mine, and has about the same thickness and the same general characteristics. Its strike here also is north and south, and its average dip in the Newport Mine about 9° to the west.

There are very few faults or disturbances of any sort so far as the works have yet extended in either of these mines, with the single exception of the gradual diminution of the dip already noticed in the lower portion of the Eastport mine. But there is evidence that within half a mile, or less, to the west of the Newport mine there is a disturbance of considerable magnitude. For the eastern part of section 8 and the western part of section 9 are occupied by a high ridge, known as Yokam Hill, in the sides of which may be seen at various points the outcrops of a bed of coal, which is in all probability identical with the one that is worked in the mines, but which here lies at an altitude above tide-water, which at one point is over three hundred feet higher than the mouth of the Newport mine. And in Yokam Hill, also, the bed, though elevated more than three hundred feet above the horizon which it occupies in the Newport mine, still dips gently to the west. There is, therefore, in all probability, either a single fault or a series of faults of considerable magnitude in the strata between Yokam Hill and the Newport mine.

There are also scattered here and there over other portions of the peninsula indications of a probability that faults and dislocations are by no means uncommon, although very little has yet been seen of them in the workings of the mines themselves.

New Mine.

As already stated, the Eastport and the Newport mines are the only ones which have hitherto been successfully worked at Coos Bay. But at the present time, March, 1877, Mr. B. B. Jones is engaged in building a railroad, and opening a new mine at a locality on the left bank of Isthmus Slough, some five or six miles above Marshfield, and opposite the little village of Coos City, which stands on the right bank at the terminus of the wagon road coming from Roseburg over the coast range of mountains; and this mine bids fair to become of greater value hereafter than either the Eastport or the Newport. It has already been proven by a tunnel driven some five hundred or six hundred feet upon a coal-bed in the direction of about north 18° east true course, from a point in the bed of a gulch in the north-east corner of the south-east quarter of section 22 of township 26, south range 13 west.

The mouth of this tunnel is one hundred and four feet above the level of low water in Isthmus Slough, thus giving plenty of height for bunkers, dumps, etc., while a railroad of only about half a mile in length, which can be cheaply built, will convey the coal from the bunkers at the mine to deep water on the slough.

At the point where the tunnel was commenced, there was no visible outcrop whatever of the coal at the surface of the ground; but a few feet beneath the surface a dark-colored streak began to show itself, which grew rapidly thicker and purer until within about one hundred feet from the mouth of the tunnel it had developed into a bed of good coal about four feet thick. Beyond that point, so far as the tunnel has yet extended, this bed will furnish from four to four and one-half feet of clean, hard coal, of a quality as good as the best hitherto furnished by the Coos Bay mines. The strike of the bed is about north 18° east true course, and its dip is about $8\frac{1}{2}°$ to the east. It has in the middle of it a streak an inch or two in thickness of a soft clay "mining;" but the remainder of the bed is good, clean coal, which separates easily from the roof and floor; and as, furthermore, both roof and floor consist of good, solid sandstone, the coal can be mined from this bed for a little less cost than it can at either of the other mines, where the roof is not so good. So far as can be judged from surface indications, also, there is every probability that this mine will prove an extensive one, and that the quantity of coal which can be cheaply extracted from it will be large. It is pleasing to be able to add that the well known experience and ability of Mr. Jones give good assurance that this property will be prudently managed, and that in so far as coal mining at Coos Bay can be made profitable hereafter, he may be expected to make this mine pay.

Mistakes and Failures.

It is not so pleasant, however, to contemplate the wasteful expenditure of money which has been made within the last few years at other localities about Coos Bay by thoroughly incompetent, if not in some cases dishonest, parties in searching and mining for coal. It probably would not exceed the truth to assert that since the year 1870 there has been expended in this way, in the immediate vicinity of Coos Bay, an aggregate sum of not less than half a million of dollars, every dollar of which is a total loss, and at least nine-tenths of which would certainly have been saved if the advice of a competent engineer had been sought and followed.

As it has been a time-honored practice among the gold and silver bearing quartz veins of California and elsewhere, to erect a costly mill, with all its apparatus complete for crushing, amalgamating, etc., before the mine was sufficiently developed to prove whether there was anything in it worth crushing or not, so the usual practice in these experiments at Coos Bay has been to build railroads, bunkers and wharves before any coal had been found that was worth mining; and in more instances than one under circumstances where there never was any reasonable probability that any coal would be found which it would pay to mine.

It may not be amiss to briefly describe two or three of these Quixotic undertakings. The first one, and perhaps the least Quixotic of all those which have involved heavy expenditure, was in progress at the time of my first visit to Coos Bay, in August, 1872, and was known as "Hardy's Mine."

This mine was located on the north-east side of the bay, nearly opposite the village of North Bend, and on the peninsula between Haynes's Slough and Jordan's Slough. Its wharf was built at Jordan's Point. The mine was opened by a tunnel whose mouth was on the north-east quarter of the south-west quarter of section 1, of township 25 south, range 13 west, and which ran in a direction north 33° east magnetic, about eight hundred and fifty feet to where it struck the coal-bed, from which point a gangway had been driven at the time of my visit (August, 1872), about four hundred feet to the north. So far as the gangway then extended, the strike of the bed was about true north, and its dip 16° or 18° to the west. This gangway lies beneath the western slope of a ridge which runs nearly north and south across the peninsula in question, and whose crest is said to rise something over four hundred feet above low water. Along the eastern slope of the ridge the coal crops out at two or three points from fifty to one hundred feet below its crest, and in these croppings the coal appeared to be of fair quality. But in the gangway below, it was found to be soft and to crumble very badly on exposure to the weather. The bed itself is, in all probability, identical with the one in the Eastport and Newport mines, as it consists of the same three layers of coal with streaks of clay-rock between them, arranged in the same order, and preserving very nearly the same thickness, both relatively and absolutely, as at those mines. But besides the fact that the coal was soft and air-slacked very badly, the bed itself, though not troubled so far as the gangway then extended with faults of any noticeable magnitude, was, nevertheless, more or less irregular and rolling in its course, and the roof was soft and weak.

Subsequent to August, 1872, the gangway was driven some distance further to the north, and the coal was said to improve in quality as they advanced. But whether in reality it did so or not, it does not appear that the improvement in any case was sufficiently great to produce a good marketable quality of coal, and the enterprise was shortly afterwards abandoned.

The mouth of the tunnel was about eighty feet above low water. Both the tunnel and the gangway, with a view to furnishing them with double tracks throughout, were foolishly driven of twice the size which was necessary; a bunker was built at the mine capable of holding about one thousand tons of coal; a railroad was built thirty-seven hundred feet in length from this bunker to the landing at Jordan's Point, and at the landing a high and costly wharf was also built.

It is the general belief among people who are more or less familiar with the history of the matter that over one hundred and fifty thousand dollars were sunk in this experiment. Certain it is, in any case, that a large sum was expended here in outside improvements and works of a character which contributed nothing towards proving the real value of the mine, and which themselves possessed no value as soon as the mine was proven worthless.

A large sum of money has also been expended to no purpose by the "Coos Bay Union Coal Company," at a locality known as the "Utter Mine." Here, also, bunkers were erected, and a railroad several miles in length was built, with wharves, etc. The mine is situated some distance to the south of the head of Isthmus Slough. Of this property I cannot speak from personal observation, not having visited the mine myself, though I have seen some very poor coal which was said to have come from there. I am informed, however, by a gentleman who has seen the mine, and whose statements I consider reliable, that in order to reach the coal a tunnel was driven in a south-easterly direction some five hundred or six hundred feet through sandstone, and that a gangway was then driven, nearly east and west, for a distance of some seventeen hundred feet along the bed, whose dip was southerly at an angle of some $16°$ or $17°$. The bed itself was about six feet in thickness, with one streak of clay-slate five or six inches thick in the middle. The upper bench of coal was of pretty good quality, but the lower one was soft. My informant thinks there would be a fair prospect for a paying mine here were it not for the cost of transportation. He adds, however, that there is only a very short lift of coal above the level of the present gangway, and as the configuration of the surface does not admit of driving a tunnel at a lower level, it would be necessary to sink in order to get the coal.

It is only necessary to add to the above statements concerning this mine, the simple fact that its location is such that the cost of transporting its coal from the mine to navigable water would prevent its competing successfully with other mines which are much more favorably situated in this respect. Mr. Utter informs me that this company has shipped altogether about ten thousand tons of coal. It might be interesting to know how much this coal has *cost per ton.*

There is also another property owned by the so-called "North Pacific Coal Company," and situated at no great distance from the Utter mine, upon which considerable money has been spent, but to which the same obstacle has been fatal from the beginning, viz.: the cost of transportation.

But among all the mistakes which have been made at Coos Bay, there is none so conspicuous or so inexcusable as that involved in the opening of what is known as the "Henryville mine;" and it would be difficult to find in the annals of coal mining anywhere an instance of grosser ignorance and incapacity, combined with more reckless folly, than has been displayed in the so-called development of this pseudo-mine.

The locality is on the right bank of Isthmus Slough, some three miles above Coos City, and about at the head of any practicable navigation in the slough. It is close to the water's edge, and the situation, therefore, would have been good enough if a mine had existed there. But there was never any reasonable evidence of any probability of the existence in this property of a pound of coal which it would pay to mine. There is, however, a heavy bed, some seven or eight feet in thickness, of alternating layers of clay-slate, "bone," and other worthless materials, including some soft and dirty coal, which strikes about north and south, and dips to the east. Upon this bed, a slope was started down, not in the direction of the dip of the bed, which, as just stated, would have been about east, but obliquely to it with a course of about N. 53° E. This slope was continued down in this direction, following approximately the general course of the plane of the bed through faults and disturbances of various kinds, and with an average pitch of about 10° or 12°, for a distance of some *twelve hundred feet*, without any visible improvement whatever in the character of the materials which constitute the bed, and with no other result than that of demonstrating the fact that this bed, worthless as it is anyhow, is also considerably broken by faults and other disturbances. This was the *mine!* But this is not all. A pair of costly and very handsomely finished steam engines, said to have been originally built for use in the Palace Hotel at San Francisco, were purchased and sent up and erected there for hoisting engines. Boilers, pumps, etc., were supplied, of course. A small town was built. A bunker was erected on piles in the water near the middle of Isthmus Slough, capable of holding nearly one thousand tons of coal. A track from the mine to the top of this bunker was laid over a trestle-work, also resting on piles, some nine hundred feet in length and about fifty feet above high water in the slough. No screens were furnished either at the mine or at the bunker, it being asserted that "this coal was of such superior quality (!) that it needed no screening, the fine being as good as the coarse." Moreover, the bunker itself was divided in its interior by vertical cross-partitions running from top to bottom into ten separate compartments. The only reason which I have ever heard assigned for this last performance is that it was done for the purpose of keeping separate from each other as many different qualities of "superior" coal; all of which, it was expected, would be obtained from this same utterly worthless bed.

There is another locality upon the same property, a little over a mile north-easterly from here, where considerable money has also been foolishly spent in sinking a slope some five hundred feet in a place where there was abundance of evidence on the surface of the ground that the rocks were badly crushed and broken, as they were also found to be in the slope itself. But this is not worthy of further description.

It is stated by good authority that the amount of money which has been expended on this Henryville property alone considerably exceeds two hundred thousand dollars. Nearly the whole of this is a dead loss, and it is with equal certainty a loss which would never have happened if any competent person had been sent to examine the property in the first place.

CHAPTER III.

WASHINGTON TERRITORY.

The coal mines of Washington Territory are situated at Bellingham Bay, close to the British Columbia line, and in the vicinity of Seattle, on the eastern shore of Puget Sound.

BELLINGHAM BAY MINE.

At Bellingham Bay there is but one mine, which is owned by the Bellingham Bay Coal Company. This mine is located at the edge of the shore of Bellingham Bay. It is opened by a slope some nine hundred feet in length, which goes down on the bed in a direction N. 30° W. magnetic, with a pitch of about 35°. About one hundred feet of the length of this slope is on trestle-work, above ground. The slope is oblique to the dip, the course of which at this point is about N. 57° W., magnetic.

The bed here worked is about fourteen feet thick; but it contains so much interstratified slate and "bone" that all the coal has to be carefully sorted by hand before it is sent to market. Moreover, after several years' experience in working the whole thickness of the bed, it was at last discovered that the lower half of it is so dirty that it is better economy to leave it in the mine than it is to attempt to work it. Accordingly, for the last few years, only about seven feet in thickness of the upper part of the bed has been mined.

At several times during its history this mine has been on fire, and once or twice it has been extinguished by flooding it with sea-water from the bay.

At various levels, one above the other in this mine, four gangways have been driven to considerable distances north-easterly from the slope. To the south-west of the slope one gangway was driven a short distance only, running directly out under the waters of the bay; but the coal in this direction was found poor, and the works took fire, and this portion of the mine is now closed up.

Above the lowest gangway, running north-easterly from the bottom of the slope, a lift extending about one hundred and fifty feet up the slope has been practically exhausted, and is now abandoned and filled with water. The three gangways remaining above the top of this lift are now known as the "upper," the "middle," and the "lower" gangways, respectively. The middle gangway is about one hundred and eighty or one hundred and ninety feet up the slope above the lower one. The lower gangway is about twenty-two hundred and fifty feet in length. Its course in starting from the slope is about N. 33° E., magnetic, and the dip of the bed is here about $33\frac{1}{2}°$. But in going north-east the gangway curves gradually around to the north and west, while at the same time the dip of the bed gradually diminishes, until at its face the gangway is running about N. 21° W., magnetic, and the dip of the bed is only 11°. For a distance of about seven hundred and fifty feet from the face of this gangway back towards the slope, the coal in the lift between this and the middle gangway is all worked out. For the remaining fifteen hundred feet the coal in this lift is nearly all solid.

The middle gangway extends north-easterly from the slope about seventeen hundred feet, and the lift between it and the upper gangway is exhausted for this whole distance, with the exception of some ten or eleven rooms stretching backwards four hundred and fifty to five hundred feet from the face, which are only about half worked out.

The upper gangway was driven over two thousand feet north-easterly, and all the coal between it and the surface of the ground is exhausted.

There probably remains in this mine an available supply of coal, if no accident happens, for three or four years to come, at its present rate of production. There is some fire-damp in this bed, and it requires constant watchfulness, though naked lights are generally used.

There has been some very poor management displayed in the history of this mine, one or two rather amusing instances of which it may be well enough to specify. The water immediately in front of the mouth of the mine is shallow. It was necessary, therefore, to build a tram-way for some little distance along the shore, in order to reach a point where the water was deep enough so that it would not be necessary to build too long a wharf. This road, which is a trifle over three-fourths of a mile in length, was built with an *ascending* grade for the load, though there was already plenty of height for bunkers, so that in hauling the coal from the mine to the bunkers it has constantly required more than two mules to do the same work which would easily have been done by one if the road had had a proper grade. Also in the bunkers, which are very heavily built, the screens are very badly arranged, and the floor of the bunker has a slope of only 25°, so that it is necessary to rehandle most of the coal after it has passed the screens in order to get the bunkers filled. Again, in the construction of the wharf itself, the attempt was made, in the face of the *teredo navalis*, which is very active and destructive here, to render the wharf more durable by merely increasing the number of the unprotected piles which were driven to support it; the result being a forest of three or four times the number of piles which were of any use beneath the wharf, all of which, of course, are being as rapidly eaten away as any single pile would be.

In the country a few miles back of Bellingham Bay, and on the north side of Whatcom Lake, there has been discovered a bed of caking coal, which the Bellingham Bay Coal Company has spent, and is spending, some thousands of dollars, rather unwisely in my opinion, in attempting to develop. This bed strikes north-easterly and south-westerly, and dips at varying angles to the north-west. At one point, which I visited in August, 1876, the bed was well exposed, and showed a thickness of from seven to eight feet. A little tunnel had also been driven in upon it here for a distance of something like a hundred feet. All of the coal from this bed would coke well; but the bed was very dirty, and contained large quantities of intermingled slate and "bone." It was easy to select hand specimens here of a clean and beautiful quality of fat, caking coal, which would be very valuable if found in a regular bed of uniform and equal quality of even half 'the thickness of the bed which was here exposed; but no large quantity of any such quality of coal was visible. There is also abundant evidence that the heavy body of nearly unaltered sandstone and shales which overlies this bed has been a good deal disturbed by faults and dislocations. But this is not the worst of the indications. The coal itself is almost immediately *underlaid* by highly metamorphic talcose and argillaceous slate, traversed in all directions by little, irregular stringers of quartz, and perfectly similar in appearance and general character to vast quantities of the same material, which may be found among the auriferous slates of the western slope of the Sierra Nevada in California.

I think, therefore, that the expenditure of probably some eight or ten thousand dollars, which is being made here in driving a tunnel for the purpose of opening up this bed in depth is not very well warranted, and that there is little probability of its developing anything sufficiently regular and uniform in quantity and quality to pay for working.

TALBOT MINE.

The only coal mines yet worked to any noteworthy extent in Washington Territory besides that of Bellingham Bay, are those of the Seattle Coal and Transportation Company, the Renton Coal Company, and the Talbot Coal Company.

Both the Renton and the Talbot mines have been opened within the last three or four years, and though I visited both of them in January, 1875, I am not so well informed with reference to their later developments as with reference to those of the Seattle Coal and Transportation Company.

The opening of the Talbot mine is situated on the south-east quarter of section 19 of township 23 north, range 5 east, Willamette meridian. In January, 1875, a tunnel had been driven here some two hundred feet, starting in a north-easterly direction, and then curving around to nearly an east course.

In the face of this tunnel there was exposed a bed of coal from six to eight feet thick, of which the lower three or four feet seemed to be pretty clean and pure, though rather soft coal. The upper portion of the bed was still softer, and was also more or less impure. The bed dipped at a gentle angle, which I estimated at 7° or 8° to the south-east. The roof and floor were also soft and weak, and there was not more than twenty-five or thirty feet of cover over the coal, even at the face of the tunnel. Nevertheless, the quality of the coal was such that I then considered the prospect fair for a development of something of a mine there. Since then a good deal of additional work has been done, and the company are now shipping a reputedly good quality of of coal to San Francisco; though whether it be at a profit or at a loss, I am not informed.

Renton Mine.

The property of the Renton Coal Company is located in sections 17, 20 and 29 of township 23 north, range 5 east, Willamette meridian. The mouth of the mine is situated near the left bank of Cedar river about a mile from its mouth, and in the south-west quarter of section 17.. Two beds have been worked here to a considerable extent, both of which strike about S. 10° W. true course, and dip to the east with an average inclination of about 15°.

The upper bed is about seventeen feet in thickness, but contains some interstratified bone, there being about ten feet of good coal in the bed. The lower bed has a total thickness of about thirteen feet, of which the lower five or six feet is so bony that it is worthless, while the upper seven or eight feet is nearly all good coal. Between these two beds there is a thickness of about eighty feet of sandstone and shales, including one bed about five feet thick of impure coal. The upper bed was first opened by a level tunnel driven directly in upon its course a distance of some twelve or fifteen hundred feet, and a considerable quantity of coal was extracted from it. But after the opening of the lower bed, the upper one was abandoned, as the coal in the lower bed was found to be cleaner, and more cheaply mined than in the upper one. The lower bed was first reached by a level tunnel driven obliquely to its course through the overlying strata, a distance of nearly two hundred feet to the coal. From this point a gangway has been driven south a distance of nearly two thousand feet, and a large quantity of coal extracted. Within this two thousand feet there has been found a single fault, consisting of a large jump up to the south. But this is the only disturbance of any magnitude that has been found.

When the Renton mine was first opened in 1874, the company built about two miles of railroad from the mouth of their mine to a point of shipment upon Black river, a mile or two below the mouth of Cedar river. From this point the coal was sent down the Black and Duwamish rivers to Seattle in barges. They subsequently, in 1875, extended this line of railroad about two miles further to a point upon the Duwamish river about nine miles from Seattle. These four miles of railroad were built along the previously surveyed and partially graded line of the Seattle and Walla Walla railroad. Still later, during the year 1876, the Seattle and Walla Walla Railroad Company built and equipped the remaining nine miles of road to Seattle, so that both the Renton and Talbot Coal companies now have a continuous all rail line direct from the mouths of the mines to Seattle.

On the north-west quarter of the south-west quarter of section 20, the Renton Company has recently struck, by boring at the depth of about seventy feet beneath the surface, a bed of coal which is believed to be identical with that upon which the Talbot Company is working, and probably different from either of those already opened by the Renton Company. It is stated that the bed, as passed through in boring, was about eleven feet thick, with, as nearly as could be judged, from nine to ten feet of good workable coal.

SEATTLE COAL AND TRANSPORTATION COMPANY'S MINES.

The openings of the mines of the Seattle Coal and Transportation Company are situated about ten miles in an air line in a south-easterly direction from the town of Seattle, and near the centre of section 27 of township 24 north, range 5 east, Willamette meridian.

From the mines a railroad of three feet 'gauge extends about three miles to the top of the bluff on the eastern shore of Lake Washington. Here the loaded cars are lowered over an automatic plane about nine hundred feet in length to the edge of the lake. They are then taken on board of a barge, and towed a distance of some eight miles to the head of Foster's Bay, a little cove in the western shore of the lake. From here a railroad extends about one-third of a mile across the portage between Lake Washington and Lake Union. The cars are then taken on a steam barge, which carries them some three miles to the southern end of Lake Union, whence a railroad about three-quarters of a mile in length leads to the point of shipment at Seattle.

The town of Seattle itself is located on the eastern shore of Duwamish Bay, and about the north-west corner of section 5, township 24 north, range 4 east.

The coal field in the vicinity of Seattle was carefully and extensively studied by Mr. T. A. Blake during the spring and summer of 1868, for the benefit of certain private parties in San Francisco; and its perfect availability and great value as a source of supply for the San Francisco market were at that time fully demonstrated by the results of his observations.

But the Mt. Diablo mines were then in the freshness of their prime, and the short-sighted policy of some of the gentlemen for whom Mr. Blake was acting, led them to doubt his opinions, and to maintain in opposition to his better judgment, that it was impossible for coal from Seattle to ever compete in the San Francisco market with coal from the Mt. Diablo mines. The result was that they neglected to secure, as they might have done at that time, for a price which was a trifle in comparison with its intrinsic worth, a coal property which is already proving itself by that sort of logic which business men, who are not engineers, can understand, to be the most valuable coal mining property yet discovered in the United States west of the Rocky Mountains.

After Mr. Blake's examinations there in 1868, the Seattle coal field lay idle for about three years. But in 1871 some new parties took hold of it, and made the first real attempts to mine coal here for market. The shipments to San Francisco were in that year a little less than five thousand tons. For the succeeding three years the shipments were also small, which was due to the fact that the owners were men of small means, and were obliged to struggle against all the disadvantages of a lack of adequate capital, combined with the steady and active opposition of a powerful clique of capitalists in San Francisco. It has, therefore, been only since 1874 that they have been in a position to be able to really begin to make the market feel what the capacities of the Seattle coal field are.

Whether Seattle coal is capable of competing with Mt. Diablo coal or not, may be judged somewhat from the statistics of the last three years. In 1874, the receipts of Mt. Diablo coal at San Francisco are stated to have been two hundred and six thousand two hundred and fifty-five tons, while in 1876 they were one hundred and eight thousand and seventy-eight tons.

In 1874, the receipts of Seattle coal are given at nine thousand and twenty-seven tons, and in 1876 at ninety-five thousand three hundred and fourteen tons. These figures are taken from the San Francisco *Commercial Herald and Market Review*, whose statistics are generally the most reliable of any that are published concerning the coal trade here.

It would exceed the proposed limits of this little volume for me to give here all the details of information contained in Mr. Blake's reports and letters of 1868, which he has kindly permitted me to use; but I shall give the more important portions of them. And, in the first place, a few words with reference to the topography of the region may not be out of place. In an unpublished report, dated April, 1868, he says:

"The coal field in question, so far as hitherto proved, lies on the east side of Lake Washington, in King County, Washington Territory. This lake, which is long and narrow, has its longer axis in a general northerly and southerly direction, and occupies a depression in the hills east of the town of Seattle. A narrow strip of high land separates it from the waters of Seattle Bay (*i. e.*, Duwamish Bay).

"Immediately behind the town, the hills, covered with dense forests of spruce and cedar, rise quite rapidly until the summit of the ridge is attained, at an elevation of perhaps two hundred and fifty feet. After a series of elevations and depressions for a distance of about a mile and a half, the ground then falls rapidly to the borders of Lake Washington.

"On the eastern shore of Lake Washington, the land is high, and is cut by streams flowing from the east and discharging into the lake.

"Two of these streams are known as Coal Creek and Honey Dew Creek, the latter being the more southerly of the two. The ridge between them, commencing abruptly at the lake, rises at intervals, until at a point some three miles east it attains an elevation of between five and six hundred feet above the level of the lake. It has its lateral cañons or swales opening north and south into the valleys of Coal and Honey Dew Creeks. It is in this main ridge, striking in a direction nearly east and west, that several heavy beds of a superior quality of coal are known to exist."

He then proceeds to describe first a bed whose outcrop was visible on the *north* side of Coal Creek and close to the line between sections 22 and 27, as follows:

"On the north side of the creek, and crossing a little ravine opening into the cañon of Coal Creek, a bed of coal has been exposed for a continuous distance of thirty feet or more. This bed may be considered at least five feet in thickness, though the carbonaceous matter extends over a width of eight feet. The upper portion however is largely intermixed with shale and is valueless. The strike of the bed is about ten degrees north of west or south of east (true course) and its dip thirty-five degrees (35°) towards the north, consequently into the hill on the north side of the creek. The bed is included in a soft sandstone which disintegrates rapidly on exposure to the atmosphere. From this point the creek flows in a general westerly direction for about half a mile. It then bends to the north-west, and within a short distance cuts the line of the coal bed. At the latter point an exposure on the bank of the creek shows a bed of coal of similar character and size with the same dip and trend. It may be considered as almost certain that the bed between these points is continuous. But its position, dipping into the hill, and the fact that the vertical height of its line of outcrop is nowhere very considerable, are not in its favor. No very great quantity of coal could be taken from above the water level, and when that was reached, mining would become more expensive, as pumping and hoisting would be necessary."

He next describes three much more important outcrops situated farther up Coal Creek, as follows:

"About the central portion of section 26, three coal beds cross the stream obliquely, and are all included within a distance of about three hundred and fifty feet at right angles to their strike. These beds show a perfect parallelism in dip and strike to each other, and to the bed already noted as occurring farther down the creek and on its northern side.

"The lowest of the three beds has been opened by a tunnel driven in upon its course from the left bank of the creek a distance of one hundred feet.

"The middle bed, one hundred and fifty feet approximately above the first, is simply exposed in the face of the bank."

"The third or upper one of the three, about two hundred feet from the second, is opened by a tunnel on its course sixty feet in length.

"The following are the details of the cross-sections of these three beds so far as can be seen, beginning with the lowest:

"No. 1. Thickness nine feet arranged as follows:

	Feet.	Inches.
Coal of superior quality, free from impurities..	3	6
Clay	0	3
Coal of superior quality, free from impurities..	3	0
Clay	0	3
Coal of superior quality, free from impurities..	2	0

"No. 2. Thickness about eight feet. Section not well exposed.

"No. 3. Thickness seven feet, arranged as follows:
Compact coal of superior quality...........4 ft. 0 in.
Clay ....0 ft. 3 in.
Coal................................8 in. to 12 in.
Clay................................3 in. to 5 in.
Coal...................................4 in.
Soft clay-slate with intermingled coal...12 in. to 15 in.

"It thus appears that the lower bed will furnish eight and one half feet of good marketable coal in a thickness of nine feet. In the middle bed the quantity is uncertain. In the upper bed the lower five feet is coal of nearly uniform quality, excepting a single clay seam of about three inches thickness, and would all be mined.

"The lower one *alone* of these three beds will probably furnish a greater mass of *good coal* in a given length and breadth than any mine yet worked on the Pacific Coast of North America.

"Proceeding westerly on the course of the beds, the hill rises quite rapidly, and the highest summit is perhaps two hundred and fifty feet above the creek where the beds have been opened by the tunnels already mentioned. Their strike is such as to continue on in the main ridge towards the lake. As an evidence of their continuity, two of them have been exposed near the east line of section 27, at a point over half a mile west from that at which they cross Coal Creek. We also have in the upper bed of all on the north side of Coal Creek, evidence of continuity of these beds, which show at these various points an identical dip and strike, and further still a probability of their extending well on towards if not to the shore of Lake Washington."

Mr. Blake adds further in this report: "From what has been said, it would appear that the property as a whole is one whose development will form an important step in the history of our productive interests. It is somewhat inconveniently located, demanding both water and land transportation before the delivery of the coal at a favorable shipping-point. However, its favorable situation and surroundings for inexpensive *mining*, together with the superior quality and great quantity of the coal, more than compensate for the other disadvantages of its location.

"The question of our future coal supply is a most important one. California, though rich in metals and abounding in wealth, is without any extensive coal field. The Mt. Diablo mines are not 'inexhaustible, and furnish but an inferior article of coal."

As already stated, this report was written in April, 1868. In September of the same year, I collated from letters received from Mr. Blake in the meantime a quantity of additional information and discoveries relating to the Seattle coal beds, from which I now extract the following:

"In May last, or about a month subsequent to the

date of the preceding report, Mr. Blake again went to Washington Territory, with the purpose, among other things, of investigating more fully than previous opportunities had permitted him to do, certain important questions relating to the extent and general character of the Seattle coal field. He has now been for some time in the vicinity of Seattle, and the following statements are collated from his recent letters."

Country West of Lake Washington.

"The strike of the coal beds, as shown upon sections 26 and 27, is about N. 80° W. (true course), and their dip is about 35° to the north.

"By reference to a map, it will be seen that the direction of this strike produced to the west, crosses Mercer Island in Lake Washington, and cuts the mouth of the Duwamish river about three miles south of Seattle.

"Moreover, the parallelism of the different beds and the apparent regularity and continuity of the inclosing strata, wherever they were exposed and could be examined, to the east of the lake, were so complete as to lend plausibility to the supposition that the coal might continue uninterruptedly westward, not only to the lake, but also across it and through Mercer Island to the opposite shore, and perhaps be ultimately again found upon the main land between the lake and the Duwamish river, or even still farther west.

"Under these circumstances, it became, of course, a matter of the greatest importance to collect all the available stratigraphical and lithological information bearing upon this question, and thus ascertain as definitely as possible the probability or improbability of such being the fact; for if the coal could be found upon or near the tide-water to the west, its extraction and shipment would become a matter involving far less expense and a corresponding increase of profit.

"To this point, then, the attention of Mr. Blake has been especially directed. But the difficulties in the way of exploration have been considerable. There are few or no exposures of the strata along the eastern shore line itself of Lake Washington, and none whatever upon Mercer Island, while the exposures upon the main land west of the lake are few and far between, the surface being generally composed of glacial drift and alluvium, and the whole country being covered with dense forests filled with fallen trees and underbrush. The investigation has therefore necessarily been slow; but with the exception of the interior of the peninsula or cape lying west of the Duwamish river, the whole country has been pretty well examined from Seattle as far south as the south end of Lake Washington, and from the eastern shore of the lake as far west as Admiralty Inlet and the bluffs along its western shore. The results of this examination, so far, have not been favorable.

"Carbonaceous matter has been discovered at various localities, but nowhere in quality and quantity sufficient to warrant the conclusion that it forms the outcrop of any workable and valuable bed of coal.

"With reference to the important question of the course and continuity of the strata, enough has been learned to show that the regularity and uniformity, which seem to be so complete throughout the coal field to the east of Lake Washington, do not continue through the country lying west of it. On the contrary, between the lake and the western shore of Admiralty Inlet, the rocks, where exposed, vary considerably in lithological character, and dip and strike in different directions, showing that there exists here either a system of varied flexures in the strata, or a series of extensive faults and dislocations, perhaps both.

"It does not necessarily follow from this that workable beds of coal may not exist somewhere in this region west of the lake. But it does follow that it is impossible to say with any certainty in advance of actual discovery, whether such is or is not the case, or to predicate with any degree of probability where they may be found, in case they do exist."

East of Lake Washington.

"But if explorations generally, west of Lake Washington, have been thus unfavorable in their results, the case has been far otherwise with the additional ones which have been made towards the east.

"Here the coal-bearing strata have been found to rest upon the northern flank of a mass of metamorphic rock which appears to form a belt stretching through the country east and west a short distance to the south of the coal beds, and may have formed an anticlinal axis to the upheaval which has raised the coal to its present position. Upon it the coal-bearing strata appear to lie with remarkable uniformity and parallelism of strike and dip wherever they have been observed, and no evidence of faults or dislocations has been discovered here. But other outcrops of coal have been found, which enhance in no small degree the probable extent and value of this coal field, already apparently immense; since they go to show the existence here of other valuable beds, different from any of those referred to in the preceding report."

Some four miles east of Lake Washington there is situated another lake known as "Sammammish," or "Squak" Lake. A creek runs into this lake from the south-east. At a point in the course of this creek, close to the line between sections 33 and 34 of township 24 north, range 6 east, "a very heavy bed of coal—at least sixteen feet in thickness—is exposed at intervals for one thousand feet or more by a small stream down the side of a steep mountain. A portion of this bed contains considerable slate. But a quarter of a mile north of it there is another bed, which is *a nine-foot bed of clear, fine coal of best quality.*" "It is probable that this last is the same bed as the lowest of the three which cross Coal Creek near the centre of section 26, township 24 north, range 5 east."

"Again, on section 27, in the latter township, at several places along a small creek which runs into the little lake called 'Boren Lake,' near the south-west corner of the same section, there is to be seen the outcrop of a very massive bed of coal, which is, in places, nearly twenty feet thick, but contains a good deal of slate. This is the lowest of all the beds yet discovered here, and may be the continuation of the lower one of the two already described as seen at 'Squak.'"

Then follow the descriptions of several other croppings. One of them on Coal Creek, in the south-east part of section 26, had been recently exposed by a cross-cut in the bank bordering the stream, and showed a thickness of from seven to eight feet. The lower part of it was slaty, and the upper portion intermixed with dirt. But at least three feet of the middle of the bed was marketable coal.

Near the south-east corner of section 25 there was another outcrop, respecting the size and character of which, however, no particulars were received. A third one was near Coal Creek, and close to the centre of the north-west quarter of section 26, and was said to promise about four feet of good coal—two feet on each side of a streak of slate or "bone" about one foot in thickness.

It was stated at the same time that the two beds already mentioned (see page 113) as having been exposed near the east line of section 27, at a point over half a mile west from that at which they cross Coal Creek were believed to be the two upper ones of "the three" which cross Coal Creek near the centre of section 26, and that the upper one of the two was here only about four feet in width, while at Coal Creek it is seven feet; the lower one also, being evidently narrower than the middle one of the three beds at Coal Creek. And this was thought to indicate a possible thinning out of the beds in going west. We shall see more clearly hereafter how nearly this surmise came to the actual truth.

I continue my extracts: "A careful re-examination has also been recently made of 'the three' beds at the point where they cross Coal Creek. There is nothing new to be said respecting the upper bed here. Some opening having been made, however, upon the middle bed, which is mentioned in the report as being simply 'exposed in the face of the bank,' a view has been obtained of its cross-section. The present condition of it, however, is not such as to reveal its true character with certainty. But, as now shown, there appears to be a band of dirt and slate, from eighteen inches to two feet wide running through the middle of the bed, with about two and a half feet of good coal on each side of it. 'With regard to the lowest of the three,—the nine foot vein,' says Mr. Blake, 'I am more and more pleased with it. The whole bed can be mined; and I do not think more than three inches in the whole nine feet will be found to be unmarketable. It is certainly a noble bed. Such a small quantity of bone compared with the mass of coal in the bed is almost inappreciable, and it can only be detected in the far end of the tunnel by the closest observation.' This tunnel will be recollected is in about one hundred feet.

"The distance along the line of strike of the beds from the most westerly point, where workable coal has yet been found, to the most easterly one at 'Squak' is between five and six miles. Throughout this distance, the strike and dip of the beds, wherever shown, is uniformly the same, and the probability is that the coal is continuous.

"It now appears, moreover, than instead of four beds of coal, which were all of which we were aware at the time of writing the preceding report, there are probably no less than five or six distinct beds within a distance of about three-quarters of a mile, measured at right angles to their strike, all of which are likely to prove workable and more or less valuable, while one of them, at least, shows a thickness of about nine feet of good solid coal in what is to all practical intents and purposes a single body.

"We say that there are *probably* no less than five or six distinct beds. There *may* be more. It is true, however, that in a country so densely timbered as this, and where the exposures of the strata are so rare, excepting along the banks of the creeks, whose beds are also filled with fallen logs, etc., it is difficult to make examinations so complete as might be desired. And it is perhaps *possible* that *some* of these appearances may be due to faults or flexures causing the same beds to reappear with perhaps slightly varied character at different points not in the same line of strike. But no evidence of any such dislocations has yet been seen; while the perfect parallelism of the beds themselves, and the uniformity in strike and dip of the inclosing strata wherever seen, speak very strongly against the probability of the existence of faults of such magnitude, or flexures of such a kind, as would be necessary in order to account in this way for the positions in which the coal is seen.

"It is certain, then, that the facts at present warrant, and we believe that développements in the future will justify, the probabilities as stated above. And if this be the case, it will not be easy to over-estimate the future importance of the Seattle coal field to the commercial and productive interests of the Pacific Coast; notwithstanding the heavy outlay which will be required to open the mines upon a proper scale, and to put the coal in the market."

I have been thus copious in my extracts from these old reports and notes of 1868, for two reasons: First, because they present as clear and concise an outline of the early discoveries at the Seattle mines as it would be possible for me to compile to-day; and second, because I consider it an act of justice to Mr. Blake himself to show that he appreciated, at a time when others refused to recognize, the great value of the Seattle coal field.

I recollect that in 1868 it was Mr. Blake's private opinion that, with adequate means to handle these mines to the best advantage, it would be possible to lay the coal from them alongside the wharves in San Francisco, at a total cost for mining and transportation, which should not exceed five dollars ($5.00) per ton in gold coin. And I recollect also that this opinion was then ridiculed as the height of absurdity. I do not hesitate to say that my own opinion in 1877 is, that, with adequate means and proper management, the same thing can be accomplished at a cost of four dollars and fifty cents ($4.50) per ton in gold coin.

I proceed now to describe the later developments of the Seattle Coal and Transportation Company. In 1871, certain of the beds had been discovered close to the surface of the ground at a point considerably to the west of the most westerly point where they had yet been actually proven to exist at the time of Mr. Blake's examinations in 1868. Work was therefore commenced at a point about six hundred feet west and some two hundred and fifty feet north of the centre of section 27, by starting two tunnels in an easterly direction into the hill. These two tunnels were at about the same level, and were about ninety feet apart horizontally, each of them being driven in directly upon the course of a bed of coal. Each tunnel, therefore, answered as a gangway through which to work and extract the coal. The southern tunnel upon the lower bed was designated as "Tunnel No. 1," and the northern one upon the upper bed as "Tunnel No. 2."

When Mr. Blake again visited the property in 1873, he found "Tunnel No. 1," to be already about twenty-four hundred feet long, and "Tunnel No. 2," about sixteen hundred feet, while a third one called "No. 3," had also been started and driven a short distance upon a bed still farther north. It appears also, that by this time, some additional prospecting had been done at Coal Creek, whereby the beds or some of them were better exposed here than they were in 1868, for he now gives the following description of five beds at the creek, beginning with the lowest:

"No. 1. As nearly as could be measured, for it is not yet opened up, is fifteen feet in thickness.

"No. 2. The point of its intersection with Coal Creek was not known previous to this visit, when in company with the superintendent and foreman of the mine, the bed was found, but no measurement of it could be made on account of the overlying gravel.

"No. 3. Is opened by a tunnel driven in westerly from the creek about one hundred feet, showing a perfect bed of very compact, clean, bright coal, eleven feet in thickness.

"No. 4. Is not opened, but a section in the bank adjoining Coal Creek shows five feet of clean coal of most excellent quality.

"No. 5 is opened by a tunnel driven in as on No. 3 for a short distance westerly on its course. There are two seams, giving in the aggregate seven feet of clear coal, separated by a clay streak two to four inches thick. This streak of clay will serve to facilitate mining."

It is here evident that what he now calls "No. 3," is the same bed which in 1868 he calls the "nine-foot bed," but which it now appears will yield eleven feet of coal instead of nine. It also appears that "No. 4," and "No. 5," are the middle and upper ones of the three beds first described at this locality, but both of them looking decidedly better in 1873 than they did in 1868; while "No. 1" and "No. 2" are underlying beds further south.

He then gives the following table, showing the thickness of the coal in the beds at both ends of the property, and the probable aggregate thickness of coal in all the beds, and remarks at the same time, that it will be seen from this table that the beds diminish in thickness going west, or increase going east.

Beds.	Thickness at Intersection with Coal Creek.	Thickness at Western End of Property.
No. 1	15 feet	10 feet (?) estimated.
" 2	5 " (?)	5 "
" 3	11 "	5 " 6 inches.
" 4	5 "	3 "
" 5	7 "	5 "
Totals	43 feet.	28 feet, 6 inches.

From this table it is evident, that at the time when it was written, Mr. Blake supposed the eleven-foot bed at Coal Creek to be identical with that bed at the western end of the property, which is five and one half feet thick, and upon which "Tunnel No. 2" was driven; while the underlying five-foot bed upon which "Tunnel No. 1" was driven was supposed to be identical with the bed designated "No. 2" at Coal Creek.

Now the actual facts of the case, as subsequently proven by further underground work, are as follows: In driving the tunnels eastward into the hill, the ninety feet of level distance, involving with the dip of 38° an actual thickness of about fifty-five feet of solid sandstone between the two beds at the mouths of the tunnels, grows gradually thinner and thinner, until, at the distance of about thirty-one hundred feet, it disappears altogether, and these two coal beds unite their aggregate thickness in one magnificent bed, which thenceforth yields from ten to eleven feet of good, clean solid coal.

This, then, is the eleven-foot bed which shows on Coal Creek; and this fact, together with some other developments which have been made since 1873, render it far less probable than it then seemed, that the beds diminish much in thickness as they go west.

The direction of these tunnels is about S. $77\frac{1}{2}°$ E. (true course), and the average dip of the beds has proven to be about 38° to the north.

My own first visit to this locality was in January, 1875, which was still some time before the two beds came together in the workings. So far as the workings then extended, the bed of tunnel No. 1 preserved an average thickness of just about five feet, and that of tunnel No. 2 of about five feet and a half. No. 2 was a little the cleaner of the two beds, containing no slate or "bone" whatever, except a single little streak of soft slate an inch or two in thickness in the centre, which served admirably for a "mining." In No. 1, a little slate might be seen here and there, somewhat irregularly distributed; but its quantity was very small, and the coal itself was said to be generally a little harder than that from No. 2. The roof and floor of both the beds were of good solid sandstone, and hardly any timbering was required in the mine. In No. 1, however, the roof and floor were not quite so regular and smooth as they were in No. 2, and the roof was not so strong, being a somewhat more micaceous sandstone, and exhibiting rather more tendency to a slaty structure.

They were, both of them, beautiful beds to work. Their thickness was good; their dip was right; their roofs and floors good; there were no faults; their coal itself was good and hard and clean; there was no pumping nor hoisting, and hardly any timbering was needed. How much every one of these items means in its bearings on the unavoidable cost of mining, will be best understood by those who have had most experience in working beds of a different character, such, for example, as those of the Mt. Diablo mines. But since the union of the two in one clean, solid bed eleven feet in thickness, the conditions for cheap mining are even more favorable than they were before, and there is not, to-day, a coal mine in North America, west of the Rocky Mountains, which for cheapness of mining, can compare with this magnificent bed.

In December, 1876, the total length of the gangway here was four thousand eight hundred and seventy feet, or, in other words, it was already driven seven hundred and seventy feet beyond the point where the two beds united; and the average height of the "lift" above the level of this gangway was between five hundred and six hundred feet.

From a point in this gangway about seventy-five feet

east of where the two beds united, a tunnel has been started and already driven (in February, 1877), one hundred and fifty feet or more in a southerly direction at right angles to the strike, in order to cut two other beds, which are already known to exist in this direction. It is expected to reach the first of these two beds, called the "Clarence Bagley Vein," or the "Seventeen-foot Vein," within a distance of about three hundred and fifty feet; and the other one, known as the "Seven-foot Vein," at a point about seventy-five feet further south. Both these beds have been traced for a long distance along the hill-tops by the sinking of little pits at intervals along their line of outcrop. The "Seventeen-foot Vein" is, as its name implies, about seventeen feet in thickness, of which at least seven feet of the lower part is said to be clean and good. The upper portion of the bed contains some "bone," but not so much but what Mr. Shattuck, the manager of the mine, thinks that it will all be workable. The "Seven-foot bed" is from six and a half to seven feet thick, and is said to be all clean, hard coal, with the exception of a single streak of clay three inches thick, about two feet above the bottom of the bed.

About one hundred and fifteen feet to the north from the mouth of "Tunnel No. 2" is the mouth of "Tunnel No. 3," which has been driven east about two hundred and twenty-five feet upon a bed of coal some five or six feet in thickness, which proved, however, too dirty and slaty to pay to work. But again, about two hundred and eighty-five feet further north comes another bed known as "No. 4." Upon this bed "Tunnel No. 4" has been driven easterly some three hundred and fifty to four hundred feet, but no coal has yet been mined here. The bed, however, proves to average about four and a half feet thick, all clean, splendid coal, of a quality a little superior to that from the eleven-foot bed.

It thus appears that in the western portion of this company's property, and near the centre of section 27, there are now known to exist not less than six beds of coal, only one of which is too poor in quality to pay for working, while the remaining five will yield an aggregate thickness of certainly not less than twenty-seven or twenty-eight, and probably from thirty to thirty-five feet of good marketable coal. Two out of the five workable beds unite, moreover, in going east, to form the eleven-foot bed.

One of the chief disadvantages under which this company has been compelled to labor has, of course, been the cost of its transportation. The present route, by which their coal is transported from the mines to Seattle, was comparatively a cheap one to build, as it involves only between four and five miles of railroad altogether, the rest of the transit being by water; but it is a very expensive route to operate, and it is now proposed to do away with it.

A narrow-gauge railroad, as already described, has now been built from Seattle up the valley of the Duwamish River to the Renton mine, a distance of about thirteen miles. From the latter point, an additional five or six miles of road is all that is required to reach the mines of the Seattle Coal and Transportation Company, and thus give them an all-rail line to Seattle, around the southern end of Lake Washington, whose total length will be but three or four miles greater than is that of their present tedious and complex line. It is intended to complete this connection during the current year.

PUYALLUP CAKING COAL.

Besides the locality already described where caking coal is known to exist near Lake Whatcom, back of Bellingham Bay, there are numerous other localities in Washington Territory at some distance back from the coast, but more particularly along and in the vicinity of the Skagit and Puyallup rivers, where more or less caking coal of good quality has been found. It is also probably true, that a little anthracite has been found at several points within the Territory. Whether any anthracite exists however, which can be made commercially available, I do not know, but do not think it probable that it does.

With reference to the caking coal upon the Skagit river, I have no information worth publishing beyond the fact that some samples of it are certainly of good quality.

But along the Puyallup river and its branches, in townships 18 and 19, north, range 6 east, Willamette Meridian, there is a considerable field of caking coal, some of which at least is of most excellent quality. This field was visited in April, 1875, by Mr. T. A. Blake, who, on his return, brought with him about a ton of the coal to San Francisco. This coal was tested for blacksmithing purposes, and at the San Francisco Gas Works for gas. It proved to be a most excellent blacksmith coal, and Mr. Blake furnishes me the following result of the test at the Gas Works:

"Two hundred and fifty pounds of coal, coarse and fine without selection, yielded over ninety-two hundred cubic feet of gas of fifteen-candle power per ton, and left one hundred and eighty-two pounds of coke of finest quality. A test of the gas for sulphuretted hydrogen, with paper dipped in solution of acetate of lead, showed that it contained less sulphur without purification than the Company's gas, made from a mixture of Australian and Nanaimo coal, contained after purification."

The bed from which this coal was taken was about six feet thick and contained a streak of slate in the middle of the bed about fifteen inches thick. Other beds are reported to exist from three to fifteen feet in thickness. It is also reported that some anthracite has been found in this vicinity.

What the commercial value of this coal field may prove to be is yet a very doubtful question, for Mr. Blake states that the beds where he saw them, "will certainly be found to be badly broken up," *i. e.* with faults and other disturbances. The problem is, however, in a fair way of being solved since the Northern Pacific Railroad Company are already courageously engaged in building a branch railroad up the valley of the Puyallup river in order to reach this coal.

CHAPTER IV.

MISCELLANEOUS.

Cost of Production at Mt. Diablo Mines.

The cost of mining and transporting the Mt. Diablo coal has varied very greatly, not only between the different mines, but also at different times and under varying circumstances for the same mine. The differences in this respect have been so great, indeed, that any single statement of the actual cost for any particular mine at any definite time would be of no value whatever as an index of the cost at the same time for a different mine, or for the same mine at a different time. This fact is well illustrated by the history of the Black Diamond Company. At their mines, the monthly averages of the cost per ton for labor alone in mining the coal and putting it into the bunkers at the mines, exclusive of the cost of timber and all other supplies, have ranged at different times since 1867 from a minimum of about two dollars and thirty-seven cents to a maximum of very nearly four dollars; or say, including supplies, from about two dollars and seventy-five cents to four dollars and fifty cents, or a little more, per ton. Within the same time, the monthly averages of the cost of the railroad transportation from the mines to the landing have ranged from twenty-five or thirty cents to over one dollar per ton; while the cost also of the water transportation from the landing to San Francisco has varied between thirty-seven cents and one dollar and twenty-five cents per ton.

But then, again, these three items of cost for mining, for land transportation, and for water transportation have rarely or never reached either their maxima or their minima values simultaneously, and consequently the actual highest or lowest figures of total cost for the mining, transportation, and delivery of the Mt. Diablo coal at any particular time would not be obtained by adding together separately either the highest or the lowest of the figures given above.

This total cost, however, has varied at different times since 1866, from a minimum of about five dollars, or possibly a little less, to a maximum of somewhere between six dollars and fifty cents and seven dollars per ton. But for a general estimate of the total average cost of all the Mt. Diablo coal which has ever been sent to market, the sum of five dollars and seventy-five dollars per ton may be taken as a fair approximation.

It may also be stated in this connection, that the average loss of coal in the pillars and in waste of one kind and another in the working of the Mt. Diablo mines has been, as nearly as it can be estimated, not far from twenty-five per cent. In other words, only about three-fourths of the coal which the beds contained has been extracted and utilized throughout the whole extent of the works.

Statistics of Production and Trade.

In order to show as nearly as may be the growth and magnitude of the coal production and trade of the Pacific Coast up to the present time, I first present the following table which is here reprinted from the columns of the San Francisco *Commercial Herald and Market Review* for January 18, 1877, without further change than the omission of some insignificant items from Queen Charlotte's Island, Sitka, Saghalien, Fuca Straits and Japan, which aggregate altogether only fifteen hundred and sixty-four tons:

ANNUAL RECEIPTS OF COAL AT SAN FRANCISCO.

Years	Mount Diablo, tons.	Coos Bay, tons.	Bellingham Bay, tons.	Vancouver Isl'd, tons.	Chile, tons.	Australia, tons.	English, tons.	Cumberland, tons.	Anthracite, tons.	Seattle, tons.	Rocky Mountain, tons.	Total tons.
1860	......	3,145	5,490	6,655	1,900	7,850	6,640	5,970	39,985	......	......	77,635
1861	6,620	4,630	10,055	6,475	12,495	23,370	23,565	2,975	26,060	......	......	116,245
1862	23,400	2,815	10,050	8,870	5,110	12,590	16,055	4,970	36,685	......	......	120,545
1863	43,200	1,185	7,750	5,745	1,790	16,890	14,660	5,670	38,660	......	......	135,550
1864	50,700	1,200	11,845	12,785	2,323	21,160	18,330	7,275	41,680	......	......	167,298
1865	60,530	1,506	14,446	18,181	1,410	17,610	9,655	4,230	22,585	......	......	150,147
1866	84,020	2,120	11,380	10,852	1,480	53,700	7,400	9,524	12,124	......	......	192,601
1867	109,490	5,415	8,899	14,829	14,949	26,619	7,302	12,177	48,518	......	......	248,925
1868	132,537	10,524	13,866	23,348	8,511	31,590	29,561	2,292	29,592	......	......	282,025
1869	148,722	14,824	20,552	14,880	1,114	75,115	17,386	11,536	24,844	......	......	326,973
1870	129,761	20,567	14,355	12,640	7,350	83,982	31,196	9,322	21,320	......	......	320,493
1871	133,485	28,690	20,284	15,621	4,161	38,942	54,191	6,060	7,231	4,918	1,025	315,194
1872	177,232	32,562	4,100	26,008	3,682	115,332	29,190	10,051	19,618	14,830	1,862	434,467
1873	171,741	38,066	21,211	31,435	400	96,435	52,616	8,857	18,295	13,572	1,904	454,582
1874	206,255	44,857	13,685	51,017	......	139,109	37,826	15,475	14,263	9,027	433	531,947
1875	142,808	32,869	10,445	61,072	......	136,869	57,849	10,324	18,810	67,106	53	538,209
1876	108,078	41,286	21,335	100,965	3,150	131,695	121,948	12,529	11,871	95,314	226	648,388

This table requires, however, a few explanatory remarks. In the first place, with reference to all the coal which comes here by sea from outside the Golden Gate, *i. e.*, *to all the coal which arrives here, except the Mt. Diablo and the Rocky Mountains,* the figures in this table have been generally obtained by taking the reports of the vessels on their arrival, and before discharging as to the quantity of coal they had on board; and these reports vary slightly in almost every cargo from the amount as actually weighed when the vessel comes to be discharged. These differences are of course small, and are sometimes in one direction and sometimes in the other. But on the average, and in the long run, it is probable that the first reports are slightly in excess of the actual quantities as weighed.

In the second place, it will be noticed that this table purports to give only the "*receipts of coal at San Francisco,*" and this is what it actually does give with a good degree of accuracy for all the other coals excepting that from the Mt. Diablo mines.

But the figures which it gives for these mines do not represent either the actual " receipts at San Francisco," or the total product of the mines. What do they represent, with a fair approach to accuracy, is *the total quantity which has been shipped away from the mines.* There is, and always has been, a large proportion of the product of the Mt. Diablo mines which has been delivered on board of steamers and other vessels at Pittsburg and New York landings, and which has been been partly burned in steamers on the bay and rivers, and partly sent direct to Sacramento, Stockton, Vallejo, and other places, without ever coming to San Francisco;

and all the coal so disposed of is included in those figures. But, on the other hand, they do not include the large item of consumption at and in the immediate vicinity of the mines themselves.

With reference to the mines of Oregon and Washington Territory, the figures in this table, being the receipts at San Francisco, represent pretty nearly nine-tenths of the total production of the mines, the aggregate consumption at and in the vicinity of the mines, and also upon ocean steamers, being not far from ten per cent. of the production.

In the case of the Vancouver Island mines, the figures probably do not represent quite so large a proportion as nine-tenths of the production; for, besides the town of Victoria and some smaller settlements, which draw their supplies almost entirely from these mines, the quantity of Vancouver Island coal which has been burned on ocean steamers is considerably larger than of Washington Territory, or of Oregon coal.

With these explanations, the above table may be taken, except in the case of the Mt. Diablo mines, for as good a general exhibit of the statistics of the coal production and coal trade of the Pacific Coast from 1860 to 1876, inclusive, as it is possible now to compile.

But, with reference to the Mt. Diablo mines, having had better facilities than any mere statistician has had for knowing the truth about the mines and their operations, I have compiled the following table, which, with the accompanying explanations and remarks, may be relied upon as furnishing a more accurate statement, and a closer approximation to the total production of these mines than has ever yet been published, or than

136   COAL MINES.

is likely to be compiled or published hereafter, for the first sixteen years of their existence.

It should be mentioned that in this table, as well as in the preceding one, all the figures are in tons of two thousand two hundred and forty pounds avoirdupois:

Years.	Black Diamond.	Union.	Pittsburg.	Eureka.	Independent.	Manhattan.	Central (approximate only).	Estimated additional product.	Total production.
1861	1,370							5,250	6,620
1862	10,672							12,728	23,400
1863	14,232							28,968	43,200
1864	12,421							38,279	50,700
1865	14,491	11,187						34,852	60,530
1866	16,009	14,224	9,599	7,391	15,678	65		21,054	84,020
1867	38,368	24,167	21,909	10,908	14,338		3,000	12,000	124,690
1868	70,100	21,641	22,920	15,815			3,001	10,200	143,676
1869	79,548	17,274	27,938	16,945			4,729	10,800	157,234
1870	70,608	20,563	23,958	10,246			5,055	11,400	141,890
1871	75,536	17,209	22,339	18,194			7,215	12,000	152,493
1872	103,008	21,494	26,714	16,831			9,612	13,200	190,859
1873	104,596	22,600	32,362	4,075			8,578	14,400	186,611
1874	117,804	30,002	43,546				9,000	15,000	215,352
1875	83,645	26.365	33,628				8,000	15,000	166,638
1876	63,048	24,000	21,801				3,000	16,200	128,049
	875,516	250,726	286,714	100,405	30,016	65	61,189	271,331	1,875,962

In the column here headed "Black Diamond" there is given, from the books of the Black Diamond Coal Company, the total production of their mines, excepting the amount consumed for hoisting and pumping under the boilers at the mines. It includes the local sales by the superintendents at the mines and at the landing, and the consumption by the locomotives on their railroad, as well as by their tug upon the river and bay.

The rest of the mines named in this table, with the exception of the "Central," are all located at Somersville.

The first shipments of coal from the Mt. Diablo mines were in 1861, and besides the old companies at Nortonville, which were all afterwards purchased by the Black Diamond Company, and whose product is therefore included in the column headed "Black Diamond," several of the Somersville companies also began to ship coal in the same year, all the coal at this time and for several years afterwards being hauled to the landings by teams. The Pittsburg railroad first began to carry coal from Somersville in March, 1866, and with the single exception of the Union Mine from the beginning of 1865, the best information which can now be obtained respecting the production of the Somersville mines prior to the completion of this railroad must be gleaned from the newspaper statistics, the old account books of these various companies, some of which were very loosely kept in the first place, having been long ago scattered about, and many of them destroyed.

In the column headed "Union," the production of that mine for the years 1865 and 1866 is given from the books of the Union Coal Company. The figures given for that mine subsequent to 1866, together with all the figures given for the Pittsburg, the Eureka, the Independent and the Manhattan Coal Companies, are from the books of the Pittsburg Railroad Company, and show the quantities transported over the railroad from the different mines respectively. These quantities, however, do not represent the total production of the mines inasmuch as they do not include, first, the local sales at Somersville and at Pittsburg Landing; second, the quantity consumed by the locomotives on the Pittsburg railroad, and third, the consumption under the boilers for pumping and hoisting at the various mines of Somersville. The Manhattan Company shipped no coal after 1866; the Independent Company shipped none after 1867; and the Eureka Company none after 1873. The Union mine was also closed and abandoned about the first of December, 1876. The Central mine stopped work in the early part of the year 1876.

In the column headed "Central (approximate only)," there is given, as nearly as it can now be ascertained, the product of the Central mine. The figures in this column for the years 1869 and 1870 are accurate and from the books. For the remaining years they are estimates based upon the best information obtainable, and are "approximate only."

The figures in the column headed "Estimated Additional Product" are estimates intended to cover the following items: First. The total production from 1861 to 1864, inclusive, of all the mines except the Black Diamond, together with the local sales and consumption under boilers for those years at the mines of that company. Second. For 1865, the total production of all the mines except the Black Diamond and the Union, together with the local sales and consumption at the mines of those two companies. Third. For 1866, the quantity hauled by teams in the first three months of that year from all the Somersville mines except the Union, the local sales at Somersville, the consumption by the locomotives of the Pittsburg railroad, and the consumption under the boilers at all the mines; also, the production in that year of the San Francisco, the Peacock, the Central, and the Teutonia mines. Fourth. From 1867 to 1876, inclusive, the consumption under boilers at all the mines, the local sales at Somersville and Pittsburg Landing, the consumption by locomotives on the Pittsburg railroad, and the total product of all mines other than those specified in the table.

It will be noticed that the figures in this column from 1861 to 1866, inclusive, are such as to make the total production for those years equal to the amounts given for the same years in the table of the *Commercial Herald and Market Review*. I have made them so, because I am disposed to believe that for these six years during which no very accurate accounts were kept, the figures in that table, though purporting to show only the "receipts at San Francisco," are, nevertheless, in all probability, large enough to cover the whole product of the mines.

In the estimate of sixteen thousand two hundred tons for 1876, there is included the product of the Empire mine for that year, which, I am informed by one of its owners, was about three thousand tons. With this single exception, more than nine-tenths of all the quantities given in this column of estimates for the ten years subsequent to 1866, were burned under the boilers at the mines for pumping and hoisting purposes, the items of local sales and consumption on the railroad being comparatively very small. The estimates are based upon a good general knowledge of the character and comparative magnitude of the operations at the different mines, and upon the fact that for several years past, although no accurate account of it has been kept, the consumption beneath the boilers at the mines of the Black Diamond Company alone is known to have averaged not far from six hundred tons per month.

Relative Values of Different Coals.

As the proximate analysis of a coal does not give the means of computing its calorific power, and as it furnishes at best but an imperfect means of estimating its practical value, I have not thought it worth while to reproduce here a table of hitherto published analyses of Pacific Coast coals. Those who are interested in these analyses will find them in the State Geological Survey Report — *Geology of California*, vol. 1, p. 30, and in a table compiled by Mr. Archibald R. Marvine, in the Annual Report for 1873, of the United States Geological and Geographical Survey of the Territories, by F. V. Hayden, pp. 113, 114. I will only present here two hitherto unpublished proximate analyses of Seattle coal, of which No. 1 was made for Goodyear & Blake, by Falkenau & Hanks, in April, 1868; and No. 2 has been furnished me by the President of the Seattle Coal and Transportation Company, and was made by Mr. H. G. Hanks, in May, 1875. They are as follows:

No.	Water.	Combustible bituminous substances.	Fixed Carbon.	Ash.	Sulphur.	Total.
1	11.66	35.49	45.98	6.44	0.43	100.00
2	6.70	38.32	47.99	6.49		99.50

Believing, however, that the results of careful working experiments upon a large scale, with reference to the relative practical values of the various coals which come to this market for steam purposes, would possess no little general interest and value, I have endeavored to gather as much reliable information of this kind as it was possible for me to obtain. In this direction I have not succeeded so well as I could wish. But I present the best information which I have, not because it is satisfactory, for it is not, but because it is all which I have been able to obtain, and because I believe that such as it is, and being reliable so far as it goes, it will not be without interest.

The most comprehensive information which I have upon this subject is embodied in the following table furnished me by the courtesy of Mr. Charles Elliot, the City Superintendent of the Spring Valley Water Works, and giving the results of a series of experiments made at the pumping works of the Spring Valley Water Company, under his supervision at various times, extending over a period of between seven and eight years.

In this table, the first column shows the kind of coal employed. The second column shows the date, *i. e.*, the month and year, and in a few cases the day of the month of the experiment. The third column shows the duration of the experiment in all cases where such duration was noted; where it was not noted, the duration was in most cases a single day. The fourth column shows the "duty" performed; *i. e.*, the number of foot-pounds of useful mechanical effect produced by each hundred pounds of coal; or, in other words, as stated at the head of the column, the number of pounds of water raised one foot high by the combustion of each hundred pounds of coal:

KIND OF COAL.	Date of Experiment.	Duration of Experiment.	"Duty," i.e., No. of lbs. raised 1 foot high by each 100 lbs. of coal.
Mt. Diablo (Eureka) Screenings.	June, 1869	....	23,678,000
Nanaimo Coal (V. I.)............	July, 1869	....	32,317,600
Mt. Diablo (Pittsburg	Feb. 1870	....	24,850,450
Anthracite (Philadelphia)	Feb. 1870	....	37,600,000
Sydney Coal (Australia)	May, 1870	....	40,032,000
" " "	June, 1870	....	36,350,000
" " "	August, 1870	....	37,036,184
Mt. Diablo (Union) Screenings.	Sept. 1870	....	25,588,636
" " " "	Sept. 1870	....	26,333,557
Anthracite......	Nov. 1870	...	46,657,500
Mt. Diablo (Black Diamond)...	Dec. 1870	1 week	25,754,400
" " (Union) Screenings.	Jan. 1871	24 hours	28,102,173
" " Screenings.	May, 1872	24 hours	23,000,000
Bellingham Bay, Screenings....	June, 1873	23 hours	29,018,600
Sydney Coal (Australia)	June, 1873	....	38,215,700
Seattle Coal..................	June,. 1873	....	29,630,000
Sydney Coal (Australia)........	Nov. 1873	1 month	36,660,000
Welsh Coal....................	Dec. 1873	24 hours	40,880,000
" "	Dec. 1873	1 day	37,252,000
" "	April, 1874	....	34,300,000
Sydney Coal (Australia)	June, 1874	30 days	38,000,000
" " "	July, 1874	7 days	38,889,200
" " "	Feb. 1875	7 days	38,681,250
Mt. Diablo (Black Diamond)...	June, 1876	1 week	25,120,000
Welsh Coal	Dec. 1876	234 hours	36,596,000

It is needless to remark upon one fact which all well-informed engineers will promptly recognize on looking over the above table, viz., that the pumping engines of the Spring Valley Water Company are very far from being up to the standard of the best pumping engines of the present day, so long as they yield less than forty-one million foot-pounds of useful effect for each hundred pounds of good anthracite coal.

But there is valuable information in the above table; and the experiments which it shows are, in spite of some rather wide variations, none the less valuable because of the internal evidence which they bear of being a true record of the best results actually obtained under the existing circumstances.

It will be seen that the "duty" of the same kind of coal varied largely at different times and in different experiments; that of the Sidney coal ranging from thirty-six millions three hundred and fifty thousand to forty millions and thirty-two thousand, and that of the Mt. Diablo from twenty-three millions to twenty-eight millions one hundred and two thousand one hundred and seventy-three foot-pounds. It is safe to assume in general that the shorter the duration of the experiment and the fewer the number of experiments with any given kind of coal, the less reliable will be the results respecting that coal. But the variations in this table are such as cannot be satisfactorily accounted for by differences in the duration of the experiments only. For instance, of two experiments with Sydney coal, each of which extended over one month's time, one gave a duty of thirty-six millions six hundred and sixty thousand, and the other a duty of thirty-eight million foot-pounds. It is therefore evident, either that the actual quality of the same denomination of coal varied considerably in the different experiments, or else, as in the light of the two consecutive experiments of December first and second, 1873, with Welsh coal, seems not improbable, that there was some irregularity in the performance of the engines themselves, which was due to causes that are not explained by the table.

If, now, without regard to the duration of the separate experiments which is stated in only twelve out of the twenty-five experiments given in the table, we take for each coal the sum of the duties, as given in the table for all the experiments with that particular kind of coal, and dividing this sum by the number of experiments, thus obtain a mean value for the duty of each of the different kinds of coal; if, then, we compare these mean values with each other, assuming for purposes of comparison that the value of the Mt. Diablo coal is unity, and stating the values of the others in unity and decimals, we obtain the following relative values for the various coals included in the above table:

Mt. Diablo Coal (Screenings)	1.000
Seattle Coal	1.171
Sydney Coal	1.502
Welsh Coal	1.472
Bellingham Bay Coal (Screenings)	1.148
Nanaimo Coal	1.277
Anthracite	1.546

In this statement of the relative values of the different coals, the figures which relate to the Mt. Diablo and the Sydney are evidently the most reliable, as the experiments with these two varieties were the most numerous, there having been eight experiments with each. Next in order of reliability comes the Welsh coal with four experiments, then the anthracite with two, and finally the Seattle, the Bellingham Bay and the Nanaimo with only one experiment each.

In addition to the preceding, Mr. Elliot has also furnished the following results of some very recent trials between the Seattle (W. T.) and the Wellington (Vancouver's Island) coals at the same works. These experiments consist of five days' run with each of the two coals. The results could not be determined in foot-pounds, for the reason that the pumps were working under somewhat variable conditions of pressure-head, etc. For the same reason, the results of a comparison between any two single days' works only would not be very reliable. But the comparison of the means for the whole five days gives probably a very fair result. The experiments were as follows:

First, with Wellington coal, at $6.50 per ton:

11	hours	run,	cost............	$5.25
12	"	"	"	5.80
14	"	"	"	6.38
11	"	"	"	4.51
11	"	"	"	5.80

Second, with Seattle coal, at $6.50 per ton:

14	hours	run,	cost..............	$8.00
13½	"	"	"	7.40
14	"	"	"	8.12
13	"	"	"	7.45
13½	"	"	"	7.45

It appears from this that with the Wellington coal the pumps ran fifty-nine hours, at a total cost of twenty-seven dollars and seventy-four cents, or an average cost of 47.017 cents per hour for coal, while with the Seattle coal they ran sixty-eight hours, at a total cost of thirty-eight dollars and forty-two cents, or an average of 56.500 cents per hour.

This shows a relative difference in value between these two coals of about twenty per cent. in favor of the Wellington over that of the Seattle coal. Or, if the Mt. Diablo coal be considered as unity, the Seattle being 1.171, then the Wellington will be 1.407.

The following experiments made under the steam-boilers at the foundry of W. T. Garratt, in July, 1876, by Mr. H. M. McCartney, for the Seattle Coal and Transportation Company, have been kindly furnished me:

TRIAL OF COAL AT GARRATT'S FOUNDRY, SAN FRANCISCO.

July 27th, 29th and 31st, 1876.

COAL.	Run.	Total coal used.	Total water used.		Total Ash.	Av. Coal per hour.	Av. Water per hour.		Amt. of coal required to evaporate 1 gal. or 1 lb. of water.		Amt. of water evaporat'd by 1 lb. of coal.		Per cent. of ash in coal.
	hours.	lbs.	gals.	lbs.	lbs.	lbs.	gals.	lbs.	lbs.	lbs.	gal.	lbs.	
Wellington..	9	1581	1175	9778	135	175.67	130.56	1086	1.346	.162	.743	6.185	8.54
Nanaimo ...	8¾	1576	1208	10053	164	185.41	142.12	1183	1.304	.157	.766	6.379	10.41
Seattle......	9	1765	1186	9870	196	196.11	131.78	1097	1.488	.179	.672	5.592	11.11

The time given in the "run" does not include the noon hour, but only the time during which the machinery was in motion. The Wellington coal made no clinkers to speak of, the Seattle very few, and the Nanaimo most of all. The latter was the only one which ran together and "caked" on the grates.

From these experiments, the Nanaimo coal would appear to be rather better than the Wellington, and if we still suppose the Mt. Diablo to be unity and the Seattle to be 1.171, we shall now find the Nanaimo to be 1.335, and the Wellington 1.295.

A comparative trial was made in December, 1874, as I am informed by the President of the Seattle Coal and Transportation Company, on one of the largest ferry boats in the Bay of San Francisco, between Seattle and Mt. Diablo coal, with the following result: The boat first ran fourteen days with Mt. Diablo coal, of which it consumed in that time three hundred and seven thousand and three hundred and eighty pounds. She then ran fourteen days with Seattle coal, doing the same work as before, with a consumption of two hundred and sixty-one thousand two hundred and eleven pounds. According to this test, the value of the Seattle coal, that of the Mt. Diablo coal being unity, is 1.177, a result which agrees very closely with that obtained for these two coals from the experiments of Mr. Elliot.

The foregoing are all the definite results of comparative experiments of this kind upon any considerable practical working scale which I have been able to obtain.

It is well known that with certain coals, the Central Pacific Railroad Company has made such experiments with care, and upon an extended scale, for its own benefit, upon its locomotives as well as upon its steamboats. But I regret to say, that upon applying to the company for the definite results of these experiments, with the permission to make them public, I met with a polite but positive refusal, upon the ground that, as this company is the largest single purchaser of coal upon this coast, they did not deem it right for them to place upon record any tests or experiments from which, perhaps, a standard might be established to the detriment of some and the benefit of others who are dealers in coal.

I confess that I am not able myself to understand the full force of this objection, well knowing, as I do, the fact that all the heavier dealers in coal in San Francisco already know the relative values of the different coals for steam, with a sufficiently close approximation to the truth to guide their action in the matter of prices, or of anything else relating to the market, as fully and as surely as any mere publication of the exact figures could do it.

But, though I could obtain no definite information from the railroad company itself, I may state that I have good reason to believe that some of their recent experiments with Seattle coal on locomotives have shown a difference, as between it and the Mt. Diablo, of over thirty per cent. in favor of the Seattle coal. How reliable these experiments may be, of course I do not know; but if reliable tests have furnished this result, then from the results already given of the tests at the Spring Valley Water Works and on the ferry boat, it would seem to follow, either that the Seattle coal compares more favorably with Mt. Diablo for locomotive use than it does for use under stationary boilers and on steamboats, or else that there has been within the last two years a very marked improvement in the quality of the coal furnished to this market from the Seattle mines. It is claimed by the owners of the mines that the latter is the fact; and it is worth noticing that the two analyses above given, the one by Mr. Hanks in 1875, and the other by Falkenau and Hanks in 1868, seem to add probability to this claim, as the later analysis shows only 6.70 per cent. of water against 11.66 per cent. in the earlier one.

If, now, we collect in tabular form the results of all the above experiments, we shall have the following table of relative values of different coals for steam, the value of the Mt. Diablo coal being assumed as unity:

RELATIVE VALUES OF DIFFERENT COALS FOR STEAM.

Kind of Coal.	Value.	Remarks.
Mt. Diablo	1.000	
Seattle	1.171	Experiments at Spring Valley Water Works.
Sydney	1.502	" " " "
Welsh	1.472	" " " "
Bellingham Bay.	1.148	" " " "
Nanaimo	1.277	" " " "
Anthracite	1.546	" " " "
Wellington	1.407	" " " "
Nanaimo	1.335	Experiments at Garratt's Foundry.
Wellington	1.295	" " " "
Seattle	1.177	Experiment on Ferry Boat.
Seattle	1.330	Probable results of tests on C. P. R. R.

The cause of the difference between the results obtained at the Water Works and those at Garratt's Foundry for the relative values of the Seattle, Nanaimo, and Wellington coals, I cannot explain, but merely give the figures as I obtained them.

CONCLUSION.

To him who has carefully read the foregoing pages, it will be apparent that the days of the old Mt. Diablo mines are numbered. Even within the few months which have elapsed since the preparation of this volume was begun, the operations of these mines have been considerably curtailed. At the time of the strike against a reduction of wages there in October, 1876, the Pittsburg Company ceased operations upon the Clark bed entirely, and withdrew the pump from their lowest level on that bed. Since that time their mining has been confined entirely to the "Little Vein," in the old Eureka ground, and to the Black Diamond bed. It is not unlikely that they may hereafter resume their work upon the Clark bed for a sufficient length of time to enable them to extract the coal which yet remains above their present lowest level. But it is not probable that they will ever sink their works any deeper upon this bed.

At about the first of December, 1876, the Union mine was finally closed, its pumps and machinery taken out, and the working of the mine entirely abandoned. It is not probable that the Union Company will ever resume work.

Of the old companies, therefore, there now remain actually at work only the Pittsburg and the Black Diamond companies. The mines of the Black Diamond

Company are in much better condition, generally, than that of the Pittsburg, and will undoubtedly hold out considerably longer, a fact which is largely due to the sound management of their able mining superintendent, Mr. Morgan Morgans. But, in the face of their necessarily heavy and constantly increasing costs of mining, they too must, ere many years, succumb to the better quality, and eventually the lower costs of production and transportation of the coals of Washington Territory and British Columbia.

Whether the hitherto unworked eastern portion of the Mt. Diablo coal field can, under existing circumstances, be worked at a profit, remains to be seen. But outside of this, there is no other coal field yet known in California which gives reasonable promise of being able to compete, to any considerable extent, with the northern mines.

Neither is it probable that the mines of Coos Bay (the only ones yet worked in Oregon), will be able many years longer to continue work at a profit in the face of the Washington Territory coals. For though the distance from San Francisco to Coos Bay is only about one-half as great as it is to Puget Sound, yet the shallow and often unsafe character of the bar at Coos Bay, the small size of the vessels which can go there at all, and the uncertainties which oftentimes attend the movements of even these small vessels, are such that the rates of freight from Coos Bay have generally ranged as high, and have often been actually higher than they were from Seattle; while it is more than probable that a company which owned and ran its own suitable steam colliers could transport coal from Seattle to San Francisco at a considerably lower cost per ton than they could do from Coos Bay. Moreover, the cost of mining at Coos Bay is greater than it is at Seattle; while at the same time the quality of the Coos Bay coal, for domestic purposes as well as for steam, is decidedly inferior to that of the more northern coals.

It is unquestionably to the mines of Washington Territory and of British Columbia, that this Pacific Coast must look hereafter, both for its chief domestic and its nearest and most reliable foreign supplies of that indispensable necessity of all civilized communities—a good article of coal.

www.ingramcontent.com/pod-product-compliance
Lightning Source LLC
Chambersburg PA
CBHW030334170426
43202CB00010B/1122